# Lecture Notes in Physics

## Volume 827



For further volumes:
http://www.springer.com/series/5304

# The Lecture Notes in Physics

The series Lecture Notes in Physics (LNP), founded in 1969, reports new developments in physics research and teaching—quickly and informally, but with a high quality and the explicit aim to summarize and communicate current knowledge in an accessible way. Books published in this series are conceived as bridging material between advanced graduate textbooks and the forefront of research and to serve three purposes:

- to be a compact and modern up-to-date source of reference on a well-defined topic
- to serve as an accessible introduction to the field to postgraduate students and nonspecialist researchers from related areas
- to be a source of advanced teaching material for specialized seminars, courses and schools

Both monographs and multi-author volumes will be considered for publication. Edited volumes should, however, consist of a very limited number of contributions only. Proceedings will not be considered for LNP.

Volumes published in LNP are disseminated both in print and in electronic formats, the electronic archive being available at springerlink.com. The series content is indexed, abstracted and referenced by many abstracting and information services, bibliographic networks, subscription agencies, library networks, and consortia.

Proposals should be sent to a member of the Editorial Board, or directly to the managing editor at Springer:

Christian Caron
Springer Heidelberg
Physics Editorial Department I
Tiergartenstrasse 17
69121 Heidelberg/Germany
christian.caron@springer.com

Lajos Diósi

# A Short Course in Quantum Information Theory

## An Approach From Theoretical Physics

Second Edition

Springer

Lajos Diósi
MTA Budapest
KFKI Research Institute for Particle and
Nuclear Physics (RMKI)
Konkoly Thege Miklós út 29-33
1525 Budapest
Hungary
e-mail: diosi@rmki.kfki.hu

ISSN 0075-8450 e-ISSN 1616-6361

ISBN 978-3-642-16116-2 e-ISBN 978-3-642-16117-9

DOI 10.1007/978-3-642-16117-9

Springer Heidelberg Dordrecht London New York



*Cover design:* eStudio Calamar, Berlin/Figueres

Printed on acid-free paper

Springer is part of Springer Science+Business Media (www.springer.com)

# Preface

Quantum information has become an independent fast growing research field. There are new departments and labs all around the world, devoted to particular or even complex studies of mathematics, physics, and technology of controlling quantum degrees of freedom. The promised advantage of quantum technologies has obviously electrified the field which had been considered a bit marginal until quite recently. Before, many foundational quantum features had never been tested or used on single quantum systems but on ensembles of them. Illustrations of reduction, decay, or recurrence of quantum superposition on single states went to the pages of regular text-books, without ever being experimentally tested. Nowadays, however, a youngest generation of specialists has imbibed quantum theoretical and experimental foundations "from infancy".

From 2001 on, in spring semesters I gave special courses for under- and postgraduate physicists at Eötvös University. The twelve lectures could not include all standard chapters of quantum information. My guiding principles were those of the theoretical physicist and the believer in the unity of physics. I achieved a decent balance between the core text of quantum information and the chapters that link it to the edifice of theoretical physics. Scholarly experience of the past five semesters will be utilized in this book

I suggest this thin book for all physicists, mathematicians and other people interested in universal and integrating aspects of physics. The text does not require special mathematics but the elements of complex vector space and of probability theories. People with prior studies in basic quantum mechanics make the perfect readers. For those who are prepared to spend many times more hours with quantum information studies, there are exhaustive monographs written by Preskill, by Nielsen and Chuang, or the edited one by Bouwmeester, Ekert, and Zeilinger. And for each of my readers, it is almost compulsory to find and read a second thin book "Short Course in Quantum Information, approach from experiments"...

*Acknowledgements* I benefited from conversations and/or correspondence with Jürgen Audretsch, András Bodor, Todd Brun, Tova Feldmann, Tamás Geszti, Thomas Konrad, and Tamás Kiss. I am grateful to them all for the generous help and useful remarks that served to improve my manuscript.

It is a pleasure to acknowledge financial support from the Hungarian Scientific Research Fund, Grant No. 49384.

Budapest, February 2006 Dr. Lajos Diósi

# Preface (extended 2nd edition)

Following the publisher's suggestion, I prepared this extended 2nd edition to include new parts and corrections to the 1st one. I also felt encouraged by my experience of continued teaching this course at Eötvös University, and of teaching special courses at Technion and at Durban University. The structure and contents of the volume have not changed much. However, a 12th chapter on "Qubit thermodynamics" and a related Appendix have been added. There are new sections on "Weak measurement, time-continuous measurement" in both Chaps. 2 and 4, and a summary on "Fock representation" of qubits in Chap. 5. Two new sections were added to Chap. 11, the one on "Period finding quantum algorithm" and the other on "Quantum error correction". A handful of new figures, visualizing the common "urn model" of statistics, intend to serve the reader's convenience. The 2nd edition gives me an opportunity to eliminate (hopefully most) errors or deficiencies in the 1st edition. This concerns basically Sects. 8.5 and 11.5, Sol. 4.1 in the 1st edition. Additional references bring into the reader's attention some recently published textbooks. The present volume remains a special one for it builds on the links between physics foundations and quantum information, and for its moderate length.

*Acknowledgements* I'm indebted to Ady Mann for his comments and careful reading of Chaps. 1–5, and to Tamás Geszti and Michael Revzen for their useful remarks. It is a pleasure to acknowledge financial support from the Hungarian Scientific Research Fund, Grant No. 75129.

Budapest, May 2010 Dr. Lajos Diósi

# Contents


| | | |
|---|---|---|
| 1 | **Introduction** | 1 |
| | 1.1 Introduction | 1 |
| | References | 3 |
| 2 | **Foundations of Classical Physics** | 7 |
| | 2.1 State Space, Equation of Motion | 7 |
| | 2.2 Operation, Mixing, Selection | 8 |
| | 2.3 Linearity of Non-Selective Operations | 9 |
| | 2.4 Measurements | 10 |
| | 2.4.1 Projective Measurement | 10 |
| | 2.4.2 Non-Projective Measurement | 12 |
| | 2.4.3 Weak Measurement, Time-Continuous Measurement | 12 |
| | 2.5 Composite Systems | 13 |
| | 2.6 Collective System | 15 |
| | 2.7 Two-State System (Bit) | 15 |
| | 2.8 Problems, Exercises | 16 |
| | References | 17 |
| 3 | **Semiclassical, Semi-Q-Physics** | 19 |
| | 3.1 Problems, Exercises | 20 |
| | References | 21 |
| 4 | **Foundations of Q-Physics** | 23 |
| | 4.1 State Space, Superposition, Equation of Motion | 23 |
| | 4.2 Operation, Mixing, Selection | 25 |
| | 4.3 Linearity of Non-Selective Operations | 26 |
| | 4.4 Measurements | 26 |
| | 4.4.1 Projective Measurement | 27 |
| | 4.4.2 Non-Projective Measurement | 29 |
| | 4.4.3 Weak Measurement, Time-Continuous Measurement | 30 |

4.4.4 Compatible Physical Quantities . . . . . . . . . . . . . . . 31
4.4.5 Measurement in Pure State . . . . . . . . . . . . . . . . . . 32
4.5 Composite Systems . . . . . . . . . . . . . . . . . . . . . . . . 33
4.6 Collective System . . . . . . . . . . . . . . . . . . . . . . . . 35
4.6.1 Problems, Exercises . . . . . . . . . . . . . . . . . . . . . 36
References . . . . . . . . . . . . . . . . . . . . . . . . . . . . . 36

**5 Two-State Q-System: Qubit Representations** . . . . . . . . . . 37
5.1 Computational Representation . . . . . . . . . . . . . . . . . 37
5.2 Pauli Representation . . . . . . . . . . . . . . . . . . . . . . 38
5.2.1 State Space . . . . . . . . . . . . . . . . . . . . . . . . . 38
5.2.2 Rotational Invariance . . . . . . . . . . . . . . . . . . . . 40
5.2.3 Density Matrix . . . . . . . . . . . . . . . . . . . . . . . . 40
5.2.4 Equation of Motion . . . . . . . . . . . . . . . . . . . . . 41
5.2.5 Physical Quantities, Measurement . . . . . . . . . . . . . 42
5.3 The Unknown Qubit, Alice and Bob . . . . . . . . . . . . . . . 43
5.4 Relationship of Computational and Pauli Representations . . . . 43
5.5 Fock Representation . . . . . . . . . . . . . . . . . . . . . . . 44
5.6 Problems, Exercises . . . . . . . . . . . . . . . . . . . . . . . 45

**6 One-Qubit Manipulations** . . . . . . . . . . . . . . . . . . . . 47
6.1 One-Qubit Operations . . . . . . . . . . . . . . . . . . . . . . 47
6.1.1 Logical Operations . . . . . . . . . . . . . . . . . . . . . . 47
6.1.2 Depolarization, Re-Polarization, Reflection . . . . . . . . . 49
6.2 State Preparation, Determination . . . . . . . . . . . . . . . . 50
6.2.1 Preparation of Known State, Mixing . . . . . . . . . . . . . 50
6.2.2 Ensemble Determination of Unknown State . . . . . . . . . 51
6.2.3 Single State Determination: No-Cloning . . . . . . . . . . . 52
6.2.4 Fidelity of Two States . . . . . . . . . . . . . . . . . . . . 53
6.2.5 Approximate State Determination and Cloning . . . . . . 53
6.3 Indistinguishability of Two Non-Orthogonal States . . . . . . . 54
6.3.1 Distinguishing Via Projective Measurement . . . . . . . . . 54
6.3.2 Distinguishing Via Non-Projective Measurement . . . . . 55
6.4 Applications of No-Cloning and Indistinguishability . . . . . . . 56
6.4.1 Q-banknote . . . . . . . . . . . . . . . . . . . . . . . . . . 56
6.4.2 Q-key, Q-Cryptography . . . . . . . . . . . . . . . . . . . 57
6.5 Problems, Exercises . . . . . . . . . . . . . . . . . . . . . . . 59
References . . . . . . . . . . . . . . . . . . . . . . . . . . . . . 60

**7 Composite Q-System, Pure State** . . . . . . . . . . . . . . . . 61
7.1 Bipartite Composite Systems . . . . . . . . . . . . . . . . . . 61
7.1.1 Schmidt Decomposition . . . . . . . . . . . . . . . . . . . 62
7.1.2 State Purification . . . . . . . . . . . . . . . . . . . . . . 62
7.1.3 Measure of Entanglement . . . . . . . . . . . . . . . . . . 63

7.1.4 Entanglement and Local Operations ............... 65
7.1.5 Entanglement of Two-Qubit Pure States ........... 66
7.1.6 Interchangeability of Maximal Entanglements ....... 67
7.2 Q-Correlations History ........................... 68
7.2.1 EPR, Einstein Nonlocality 1935 ................. 68
7.2.2 A Non-Existing Linear Operation 1955 ............ 69
7.2.3 Bell Nonlocality 1964 ......................... 70
7.3 Applications of Q-Correlations ...................... 73
7.3.1 Superdense Coding .......................... 73
7.3.2 Teleportation .............................. 74
7.4 Problems, Exercises ............................... 76
References ........................................ 76

**8 All Q-Operations** ...................................... 79
8.1 Completely Positive Maps ........................... 79
8.2 Reduced Dynamics ................................ 80
8.3 Indirect Measurement .............................. 81
8.4 Non-Projective Measurement Resulting from Indirect Measurement .................................. 83
8.5 Entanglement and LOCC ............................ 84
8.6 Open Q-System: Master Equation ..................... 85
8.7 Q-Channels ...................................... 85
8.8 Problems, Exercises ............................... 86
References ........................................ 87

**9 Classical Information Theory** ........................... 89
9.1 Shannon Entropy, Mathematical Properties .............. 89
9.2 Messages ........................................ 90
9.3 Data Compression ................................. 90
9.4 Mutual Information ................................ 92
9.5 Channel Capacity ................................. 93
9.6 Optimal Codes .................................... 94
9.7 Cryptography and Information Theory ................. 94
9.8 Entropically Irreversible Operations ................... 95
9.9 Problems, Exercises ............................... 96
References ........................................ 96

**10 Q-Information Theory** .................................. 97
10.1 Von Neumann Entropy, Mathematical Properties .......... 97
10.2 Messages ........................................ 98
10.3 Data Compression ................................. 99
10.4 Accessible Q-Information ........................... 101
10.5 Entanglement: The Resource of Q-Communication ......... 102
10.6 Entanglement Concentration (Distillation) ............... 103

10.7 Entangement Dilution . . . . . . . . . . . . . . . . . . . . . . . . . . . . . 104
10.8 Entropically Irreversible Operations . . . . . . . . . . . . . . . . . . . 105
10.9 Problems, Exercises . . . . . . . . . . . . . . . . . . . . . . . . . . . . . . . 106
References . . . . . . . . . . . . . . . . . . . . . . . . . . . . . . . . . . . . . . . 108

**11 Q-Computation** . . . . . . . . . . . . . . . . . . . . . . . . . . . . . . . . . 109
11.1 Parallel Q-Computing . . . . . . . . . . . . . . . . . . . . . . . . . . . . . 109
11.2 Evaluation of Arithmetic Functions. . . . . . . . . . . . . . . . . . . . 110
11.3 Oracle Problem: The First Q-Algorithm . . . . . . . . . . . . . . . . 111
11.4 Searching Q-Algorithm . . . . . . . . . . . . . . . . . . . . . . . . . . . . 113
11.5 Fourier Algorithm. . . . . . . . . . . . . . . . . . . . . . . . . . . . . . . . 115
11.6 Period Finding Q-Algorithm . . . . . . . . . . . . . . . . . . . . . . . . 116
11.7 Error Correction . . . . . . . . . . . . . . . . . . . . . . . . . . . . . . . . . 118
11.8 Q-Gates, Q-Circuits . . . . . . . . . . . . . . . . . . . . . . . . . . . . . . 120
11.9 Problems, Exercises . . . . . . . . . . . . . . . . . . . . . . . . . . . . . . 121
References . . . . . . . . . . . . . . . . . . . . . . . . . . . . . . . . . . . . . . . 122

**12 Qubit Thermodynamics** . . . . . . . . . . . . . . . . . . . . . . . . . . . . 123
12.1 Thermal Qubit . . . . . . . . . . . . . . . . . . . . . . . . . . . . . . . . . . 123
12.2 Ideal Qubit Gas . . . . . . . . . . . . . . . . . . . . . . . . . . . . . . . . . 124
12.3 Informatic and Thermodynamic Entropies . . . . . . . . . . . . . . 125
12.4 Q-Thermalization . . . . . . . . . . . . . . . . . . . . . . . . . . . . . . . . 126
12.5 Q-Refrigerator . . . . . . . . . . . . . . . . . . . . . . . . . . . . . . . . . . 127
12.6 Thermal Qubit with External Work. . . . . . . . . . . . . . . . . . . . 129
12.7 Q-Carnot Cycle . . . . . . . . . . . . . . . . . . . . . . . . . . . . . . . . . . 130
12.8 Problems, Exercises . . . . . . . . . . . . . . . . . . . . . . . . . . . . . . 132
References . . . . . . . . . . . . . . . . . . . . . . . . . . . . . . . . . . . . . . . 133

**Appendix** . . . . . . . . . . . . . . . . . . . . . . . . . . . . . . . . . . . . . . . 135

**Solutions**. . . . . . . . . . . . . . . . . . . . . . . . . . . . . . . . . . . . . . . 141

**Index** . . . . . . . . . . . . . . . . . . . . . . . . . . . . . . . . . . . . . . . . . 159

# Symbols, Acronyms, Abbreviations

| | |
|---|---|
| $\{\ ,\ \}$ | Poisson bracket |
| $[\ ,\ ]$ | Commutator |
| $\{\ ,\ \}$ | Anti-commutator |
| $\langle\ \rangle$ | Expectation value |
| $\hat{O}$ | Matrix |
| $\hat{O}^\dagger$ | Adjoint matrix |
| $x, y, \ldots$ | Phase space points |
| $\Gamma$ | Phase space |
| $\rho(x)$ | Phase space distribution, Classical state |
| $\mathrm{x}, \mathrm{y}, \ldots$ | Binary numbers |
| $\mathrm{x}_1\mathrm{x}_2 \ldots \mathrm{x}_\mathrm{n}$ | Binary string |
| $\rho(\mathrm{x})$ | Discrete classical state |
| $\mathcal{M}$ | Operation |
| $\mathcal{T}$ | Polarization reflection |
| $\mathcal{I}$ | Identity operation |
| $\mathcal{L}$ | Lindblad generator |
| $A(x), A(\mathrm{x})$ | Classical physical quantities |
| $H(x)$ | Hamilton function |
| $P$ | Indicator function |
| $\Pi(x), \Pi(\mathrm{x})$ | Classical effect |
| $\mathcal{H}$ | Hilbert space |
| $d$ | Vector space dimension |
| $\|\psi\rangle, \|\varphi\rangle, \ldots$ | State vectors |
| $\langle\psi\|, \langle\varphi\|, \ldots$ | Adjoint state vectors |
| $\langle\psi\|\varphi\rangle$ | Complex inner product |
| $\langle\psi\|\hat{O}\|\varphi\rangle$ | Matrix element |
| $\hat{\rho}$ | Density matrix, Quantum state |
| $\hat{A}$ | Quantum physical quantity |
| $\hat{H}$ | Hamiltonian |
| $\hat{P}$ | Hermitian projector |

| | |
|---|---|
| $\hat{I}$ | Unit matrix |
| $\hat{U}$ | Unitary map |
| $\hat{\Pi}$ | Quantum effect |
| $p$ | Probability |
| $w$ | Weight in mixture |
| $\vert\mathrm{x}\rangle$ | Computational basis vector |
| $\hat{\mathrm{x}}$ | Qubit Hermitian matrix |
| q- | Quantum |
| cNOT | Controlled NOT |
| $\oplus$ | Modulo sum |
| $\circ$ | Composition |
| $\times$ | Cartesian product |
| $\otimes$ | Tensor product |
| tr | Trace |
| $\mathrm{tr}_A$ | Partial trace |
| $\vert\uparrow\rangle, \vert\downarrow\rangle$ | Spin-up, spin-down basis |
| $n, m \ldots$ | Bloch unit vectors |
| $\vert n\rangle$ | Qubit state vector |
| $\boldsymbol{s}$ | Qubit polarization vector |
| $\hat{\sigma}_x, \hat{\sigma}_y, \hat{\sigma}_z$ | Pauli matrices |
| $\hat{\boldsymbol{\sigma}}$ | Vector of Pauli matrices |
| $\boldsymbol{a}, \boldsymbol{b}, \boldsymbol{\alpha}, \ldots$ | Real spatial vectors |
| $\boldsymbol{ab}$ | Real scalar product |
| $\hat{a}, \hat{a}^\dagger$ | Emission, absorption matrices |
| $X, Y, Z$ | One qubit Pauli gates |
| $H$ | Hadamard gate |
| $T(\varphi)$ | Phase gate |
| $F$ | Fidelity |
| $E$ | Entanglement measure |
| $S(\rho), S(p)$ | Shannon entropy |
| $S(\hat{\rho})$ | von Neumann entropy |
| $S(\rho'\vert\vert\rho), S(\hat{\rho}'\vert\vert\hat{\rho})$ | Relative entropy |
| $\vert\Psi^\pm\rangle, \vert\Phi^\pm\rangle$ | Bell basis vectors |
| $\hat{M}_n$ | Kraus matrices |
| $\vert n; E\rangle$ | Environmental basis vector |
| $X, Y, \ldots$ | Classical message |
| $H(X), H(Y)$ | Shannon entropy |
| $H(X\vert Y)$ | Conditional Shannon entropy |
| $I(X{:}Y)$ | Mutual information |
| $C$ | Channel capacity |
| $\rho(\mathrm{x}\vert\mathrm{y})$ | Conditional state |
| $\rho(\mathrm{y}\vert\mathrm{x})$ | Transfer function |
| $T$ | Temperature |

| | |
|---|---|
| $E$ | Energy |
| $W$ | Work |
| $Q$ | Heat |
| $S_{th}$ | Thermodynamic entropy |
| LO | Local Operation |
| LOCC | Local Operation and Classical Communication |

# Chapter 1
# Introduction

## 1.1 Introduction

Classical physics—the contrary to quantum—means all those fundamental dynamical phenomena and their theories which became known until the end of the nineteenth century, from our studying the macroscopic world. Galileo's, Newton's, and Maxwell's consecutive achievements, built one on top of the other, obtained their most compact formulation in terms of the classical canonical dynamics. At the same time, the conjecture of the atomic structure of the microworld was also conceived. By extending the classical dynamics to atomic degrees of freedom, certain microscopic phenomena also appearing at the macroscopic level could be explained correctly. This yielded indirect, yet sufficient, proof of the atomic structure. But other phenomena of the microworld (e.g., the spectral lines of atoms) resisted the natural extension of the classical theory to the microscopic degrees of freedom. After Planck, Einstein, Bohr, and Sommerfeld, there had formed a simple constrained version of the classical theory. The naively *quantized* classical dynamics was already able to describe the non-continuous (discrete) spectrum of stationary states of the microscopic degrees of freedom. But the detailed dynamics of the transitions between the stationary states was not contained in this theory. Nonetheless, the successes (e.g., the description of spectral lines) shaped already the *dichotomous* physics world concept: the microscopic degrees of freedom obey other laws than macroscopic ones do. After the achievements of Schrödinger, Heisenberg, Born, and Jordan, the *quantum theory* emerged to give the complete description of the microscopic degrees of freedom in perfect agreement with experience. This quantum theory was not a mere quantized version of the classical theory anymore. Rather it was a totally new formalism of completely different structure than the classical theory, which was applied professedly to the microscopic degrees of freedom. As for the macroscopic degrees of freedom, one continued to insist on the classical theory.

L. Diósi, *A Short Course in Quantum Information Theory*,
Lecture Notes in Physics, 827, DOI: 10.1007/978-3-642-16117-9_1,

For a sugar cube, the center of mass motion is a macroscopic degree of freedom. For an atom, it is microscopic. We must apply the classical theory to the sugar cube, and the quantum theory to the atom. Yet, there is no sharp boundary of where we must switch from one theory to the other. It is, furthermore, obvious that the center of mass motion of the sugar cube should be derivable from the center of mass motions of its atomic constituents. Hence a specific inter-dependence exists between the classical and the quantum theories, which must give consistent resolution for the above dichotomy. The von Neumann "axiomatic" formulation of the quantum theory represents, in the framework of the dichotomous physics world concept, a description of the microworld maintaining the perfect harmony with the classical theory of the macroworld.

Let us digress about a natural alternative to the dichotomous concept. According to it, all macroscopic phenomena can be reduced to a multitude of microscopic ones. Thus in this way the basic physical theory of the universe would be the quantum theory, and the classical dynamics of macroscopic phenomena should be deducible from it, as limiting case. But the current quantum theory is not capable of holding its own. It refers to genuine macroscopic systems as well, thus requiring classical physics as well. Despite the theoretical efforts in the second half of the twentieth century, so far there has not been consensus regarding the (universal) quantum theory which would in itself be valid for the whole physical world.

This is why we keep the present course of lectures within the framework of the dichotomous world concept. The "axiomatic" quantum theory of von Neumann will be used. Among the bizarre structures and features of this theory, discreteness (quantumness) was the earliest, and the theory also drew its name from it. Yet another odd prediction of quantum theory is the inherent randomness of the microworld. During the decades, further surprising features have come to light. It has become "fashion" to deduce paradoxical properties of quantum theory. There is a particular range of paradoxical predictions (Einstein–Podolsky–Rosen, Bell) which exploits such correlations between separate quantum systems which could never exist classically. Another cardinal paradox is the non-clonability of quantum states, meaning the fidelity of possible copies will be limited fundamentally and strongly.

The initial role of the paradoxes was better knowledge of quantum theory. We learned the *differenciae specificae* of the quantum systems with respect to the classical ones. The consequences of the primarily paradoxical quantumness are understood relatively well and also their advantage is appreciated with respect to classical physics (see, e.g., semiconductors, superconductivity, superfluidity). By the end of the twentieth century the paradoxes related to *quantum-correlations* have come to the front. We started to discover their advantage only in the past decade. The keyword is: *information*! Quantum correlations, consequent upon quantum theory, would largely extend the options of classical information manipulation including information storage, coding, transmitting, hiding, protecting, evaluating, as well as algorithms and game strategies. All these represent the field of quantum information theory in a wider sense. Our short course covers the basic components only, at the introductory level.

Chapters 2–4 summarize the classical, the semiclassical, and the quantum physics. The Chaps. 2 and 4 look almost like mirror images of each other. I intended to exploit the maximum of existing parallelism between the classical and quantum theories, and to isolate only the essential differences in the present context. Chapter 5 introduces the text-book theory of abstract two-state quantum systems. Chapter 6 discusses their quantum informatic manipulations and presents two applications: copy-protection of banknotes and of cryptographic keys. Chapter 7 is devoted to composite quantum systems and quantum correlations (also called entanglement). An insight into three theoretical antecedents is discussed and finally I show two quantum informatic applications: superdense coding and teleportation. Chapter 8 introduces us to the modern theory of quantum operations. The first parts of Chaps. 9 and 10 are again mirror images of each other. The foundations of classical and quantum information theories, based respectively on the Shannon and von Neumann entropies, can be displayed in parallel terms. This holds for the classical and quantum theories of data compression as well. There is, however, a separate section in Chap. 10 to deal with the entanglement as a resource, and with its conversions which all make sense only in quantum context. Chapter 11 offers simple introduction to the quintessence of quantum information which is quantum algorithms. I present the concepts that lead to the idea of the quantum computer. Two quantum algorithms will close the chapter: solution of the oracle and of the searching problems. A short section of divers Problems and Exercises follows each chapter. This can, to some extent, compensate the reader for the laconic style of the main text. A few missing or short-spoken proofs and arguments find themselves as Problems and Exercises. That gives a hint how the knowledge, comprised into the economic main text, could be derived and applied.

For further reading, we suggest the monograph by Nielsen and Chuang [1] which is the basic reference work for the time being, together with by Preskill [2] and edited by Bouwmeester, Ekert and Zeilinger [3]. Certain statements or methods, e.g. in Chaps. 10 and 11, follow [1] or [2] and can be checked there directly. Our bibliography continues with textbooks [4–10] on the traditional fields, like, e.g. the classical and quantum physics, which are necessary for quantum information studies. References to two useful reviews on q-cryptography [11] and on q-computation are also included [12]. The rest of the bibliography consists of a very modest selection of the related original publications [13–50] and of recent related textbooks [51–55].

## References

1. Nielsen, M.A., Chuang, I.L.: Quantum Computation and Quantum Information. Cambridge University Press, Cambridge (2000)
2. Preskill, J.: Quantum computation and information, (Caltech 1998). http://theory.caltech.edu/people/preskill/ph229/

3. Bouwmeester, D., Ekert, A., Zeilinger, A. (eds): The Physics of Quantum Information: Quantum Cryptography, Quantum Teleportation, Quantum Computation. Springer, Berlin (2000)
4. Landau, L.D., Lifshitz, E.M.: Course of Theoretical Physics: Mechanics. Pergamon, Oxford (1960)
5. Gutzwiller, M.C.: Chaos in Classical and Quantum Mechanics. Springer, Berlin (1991)
6. von Neumann, J.: Mathematical Foundations of Quantum Mechanics. Princeton University Press, Princeton (1955)
7. Peres, A.: Quantum Theory: Concepts and Methods. Kluwer, Dordrecht (1993)
8. Kraus, K.: States, Effects, and Operations: Fundamental Notions of Quantum Theory. Springer, Berlin (1983)
9. Busch, P., Lahti, P.J., Mittelstadt, P.: The Quantum Theory of Measurement. Springer, Berlin (1991)
10. Joos, E., Zeh, H.D., Kiefer, C., Giulini, D., Kupsch, K., Stamatescu, I.O.: Decoherence and the Appearance of a Classical World in Quantum Theory, 2nd edn. Springer, Berlin (2003)
11. Gisin, N., Ribordy, G., Tittel, W., Zbinden, H.: Rev. Mod. Phys. **74**, 145 (2002)
12. Ekert, A., Hayden, P., Inamori, H.: Basic Concepts in Quantum Computation, (Les Houches lectures 2000); Los Alamos e-print arXiv: quant-ph/0011013
13. Aharonov, Y., Albert, D.Z., Vaidman, L.: Phys. Rev. Lett. **60**,1351 (2008)
14. Diósi, L.: Weak measurements in quantum mechanics. In: Françoise, J.P., Naber, G.L., Tso, S.T. (eds) Encyclopedia of Mathematical Physics, vol. 4, pp. 276–282. Elsevier, Oxford (2006)
15. Werner, R.F.: Phys. Rev. A **40**, 4277 (1989)
16. Wootters, W.K., Zurek, W.K.: Nature **299**, 802 (1982)
17. Ivanovic, I.D.: Phys. Lett. A **123**, 257 (1987)
18. Wiesner, S.: SIGACT News **15**, 77 (1983)
19. Bennett, C.H.: Phys. Rev. Lett. **68**, 3121 (1992)
20. Bennett, C.H., Brassard, G.: Quantum cryptography: public key distribution and coin tossing, In: Proceedings of IEEE International Conference on Computers, Systems and Signal Processing, IEEE Press, New York (1984)
21. Braunstein, S.L., Mann, A., Revzen, M.: Phys. Rev. Lett. **68**, 3259 (1992)
22. Einstein, A., Podolsky, B., Rosen, N.: Phys. Rev. **47**, 777 (1935)
23. Stinespring, W.F.: Proc. Am. Math. Soc. **6**, 211 (1955)
24. Peres, A.: Phys. Rev. Lett. **77**, 1413 (1996)
25. Horodecki, M., Horodecki, P., Horodecki, R.: Phys. Lett. A **223**, 1 (1996)
26. Bell, J.S.: Physics **1**, 195 (1964)
27. Clauser, J.F., Horne, M.A., Shimony, A., Holt, R.A.: Phys. Rev. Lett. **23**, 880 (1969)
28. Popescu, S.: Phys. Rev. Lett. **74**, 2619 (1995)
29. Bennett, C.H., Wiesner, S.J.: Phys. Rev. Lett. **69**, 2881 (1992)
30. Bennett, C.H., Brassard, G., Crépeau, C., Jozsa, R., Peres, A., Wootters, W.K.: Phys. Rev. Lett. **70**, 1895 (1993)
31. Lindblad, G.: Commun. Math. Phys. **48**, 199 (1976)
32. Gorini, V., Kossakowski, A., Sudarshan, E.C.G.: J. Math. Phys. **17**, 821 (1976)
33. Shannon, C.E.: Bell Syst. Tech. J. **27**, 379–623 (1948)
34. Schumacher, B.: Phys. Rev. A **51**, 2738 (1995)
35. Holevo, A.S.: Problems Inf. Transm. **5**, 247 (1979)
36. Feynman, R.P.: Int. J. Theor. Phys. **21**, 467 (1982)
37. Deutsch, D.: Proc. R. Soc. Lond. A **400**, 97 (1985)
38. Deutsch, D., Jozsa, R.: Proc. R. Soc. Lond. A **439**, 533 (1992)
39. Grover, L.K.: A fast quantum mechanical algorithm for database search. In: Proceedings of the 28th Annual STOC, Association for Computer Machinery, New York (1996)
40. Shor, P.W.: Algorithms for quantum computation: discrete logarithm and factoring. In: 35th Annual Symposium on Foundations of Computer Science. IEEE Press, Los Alamitos (1994)
41. Shor, P.W.: Phys. Rev. A **52**, 2493 (1995)

42. Tucci, R.R.: QC Paulinesia. http://www.ar-tiste.com/PaulinesiaVer1.pdf (2004)
43. Scarani, V., Ziman, M., Štelmanovič, P., Gisin, N., Bužek, V.: Phys. Rev. Lett. **88**, 097905 (2002)
44. Linden, N., Popescu, S., Skrzypczyk, P.: How small can thermal machines be? Towards the smallest possible refrigerator, arXiv:0908.2076v1 (quant-ph) (2009)
45. Alicki, R.: J. Phys. **12**, 103 (1979)
46. Spohn, H., Lebowitz, J.L.: Adv. Chem. Phys. **38**, 109 (1979)
47. Geva, E., Kosloff, R.: J. Chem. Phys. **96**, 3054 (1992)
48. Landauer, R.: IBM J. Res. Dev. **5**, 183 (1961)
49. Diósi, L., Feldmann, T., Kosloff, R.: Int. J. Quant. Inf. **4**, 99 (2006)
50. Csiszár, I., Hiai, F., Petz, D.: J. Math. Phys. **48**, 092102 (2007)
51. Jaeger, G.: Quantum Information: An Overview. Springer, Berlin Heidelberg New York (2007)
52. Bruss, D., Leuchs G.: (editors): Lectures on Quantum Information. Wiley-VCH, Weinheim (2007)
53. Stolze, J., Suter, D.: Quantum Computing: A Short Course from Theory to Experiment. Wiley-VHC, Weinheim (2008)
54. Barnett. SM.: Quantum Information. Oxford University Press, Oxford (2009)
55. Bennett, C., DiVincenzo, D.P., Wootters, W.K.: Quantum Information Theory. Springer, Berlin Heidelberg New York (2009)

# Chapter 2
# Foundations of Classical Physics

We choose the classical canonical theory of Liouville because of the best match with the q-theory—a genuine statistical theory. Also this is why we devote the particular Sect. 2.4 to the measurement of the physical quantities. Hence the elements of the present chapter will most faithfully reappear in Chap. 4 on Foundations of q-physics. Let us observe the similarities and the differences!

## 2.1 State Space, Equation of Motion

The state space of a system with $n$ degrees of freedom is the phase space

$$\Gamma = \{(q_k, p_k); k = 1,2,\ldots,n\} \equiv \{x_k; k = 1,2,\ldots,n\} \equiv \{x\}, \tag{2.1}$$

where $q_k$, $p_k$ are the canonically conjugate coordinates of each degree of freedom in turn. The *pure* state of an individual system is described by the phase point $\bar{x}$. The generic state is *mixed* (Fig. 2.1), described by normalized distribution function

$$\rho \equiv \rho(x) \geq 0, \quad \int \rho \mathrm{d}x = 1. \tag{2.2}$$

The generic state is interpreted on the statistical ensemble of identical systems. The distribution function of a pure state reads

$$\rho_{\mathrm{pure}}(x) = \delta(x - \bar{x}). \tag{2.3}$$

Dynamical evolution of a closed system is determined by its real Hamilton function $H(x)$. The Liouville equation of motion takes this form[1]:

[1] The form $\mathrm{d}\rho/\mathrm{d}t$ is used to match the tradition of q-theory notations, c.f. Chap. 4, it stands for $\partial\rho(x,t)/\partial t$.

L. Diósi, *A Short Course in Quantum Information Theory*,
Lecture Notes in Physics, 827, DOI: 10.1007/978-3-642-16117-9_2,

**Fig. 2.1** Urn model of statistical ensemble. To visualize the statistical ensemble for a state $\rho$, we imagine an urn to contain a very large number of copies of the given classical system. The preparation of such ensemble is never unique if the state is mixed

$$\frac{\mathrm{d}}{\mathrm{d}t}\rho = \{H, \rho\}, \tag{2.4}$$

where $\{\cdot, \cdot\}$ stands for the Poisson brackets. For pure states, this yields the Hamilton equation of motion

$$\frac{\mathrm{d}\bar{x}}{\mathrm{d}t} = -\{H, \bar{x}\}; \quad H = H(\bar{x}). \tag{2.5}$$

Its solution $\bar{x}(t) \equiv U(\bar{x}(0); t)$ represents the time-dependent invertible map $U(t)$ of the state space. It enables us to construct the solution of the Liouville equation (2.4):

$$\rho(t) = \rho(0) \circ U^{-1}(t) \equiv \mathcal{M}(t)\rho(0), \tag{2.6}$$

where $\mathcal{M}(t)$ denotes the corresponding reversible operation. Below is introduced the generic notion of classical operation.

## 2.2 Operation, Mixing, Selection

Let operation $\mathcal{M}$ on a given state $\rho$ mean that we perform the same transformation on each system of the corresponding statistical ensemble. Mathematically, $\mathcal{M}$ is linear norm-preserving map of positive kernel, transforming an arbitrary state $\rho$ into a new state $\mathcal{M}\rho$.

Randomly mixing the elements of two ensembles of states $\rho_1$ and $\rho_2$ with respective rates $w_1 \geq 0$ and $w_2 \geq 0$ yields the new ensemble of state:

$$\rho = w_1\rho_1 + w_2\rho_2; \quad w_1 + w_2 = 1. \tag{2.7}$$

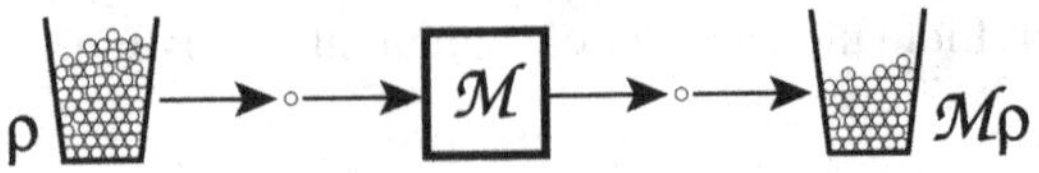


**Fig. 2.2** Urn model of operation. We draw a system of state $\rho$ from the first urn, let it go through the transformator $\mathcal{M}$ and place it into the second urn. We repeat this procedure to reach a very large number of copies in the second urn which will then represent the transformed state $\mathcal{M}\rho$

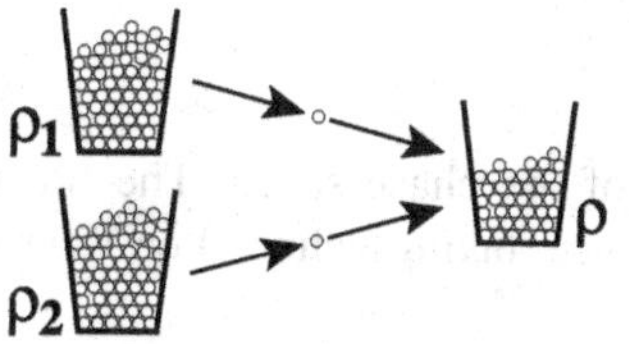


**Fig. 2.3** Urn model of mixing. We draw a system of state $\rho_1$ or $\rho_2$ from the two urns on the lhs chosen randomly at respective probabilities $w_1$ and $w_2$; we place it into the urn on the rhs. We repeat this procedure to reach a very large number of copies in the urn on the rhs, which will then represent the mixed state $\rho = w_1\rho_1 + w_2\rho_2$

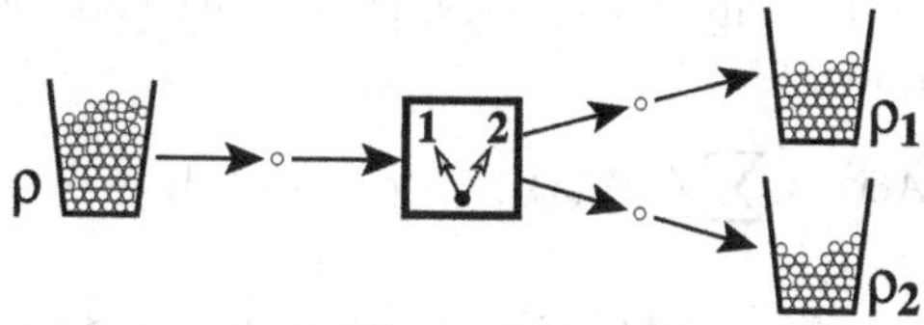


**Fig. 2.4** Urn model of selection. We draw a system of state $\rho$ from the urns on the lhs and let it go through the selective operation, e.g., a measurement apparatus; we place the system into the first or second urn on the rhs according to the outcome 1 or 2 of the measurement. We repeat this procedure to reach a very large number of copies in the urns on the rhs, which will then represent the selected states $\rho_1$ and $\rho_2$, respectively

A generic mixed state can always be prepared (i.e. decomposed) as the mixture of two or more other mixed states in infinitely many different ways. After mixing, however, it is totally impossible to distinguish in which way the mixed state was prepared. It is crucial, of course, that mixing must be probabilistic. A given mixed state can also be prepared (decomposed) as a mixture of pure states and this mixture is unique.

Selection of a given ensemble into specific sub-ensembles, a contrary process to mixing, will be possible via so-called selective operations. They correspond mathematically to norm-reducing positive maps. The most typical selective operations are called measurements, c.f. Sect. 2.4.

## 2.3 Linearity of Non-Selective Operations

The operation's categorical linearity follows from the linearity of the procedure of mixing (2.7). Obviously we must arrive at the same state if we mix two states first and then we subject the systems of the resulting ensemble to the operation $\mathcal{M}$ or, alternatively, we perform the operation prior to mixing the two ensembles together:

$$\mathcal{M}(w_1\rho_1 + w_2\rho_2) = w_1\mathcal{M}\rho_1 + w_2\mathcal{M}\rho_2. \tag{2.8}$$

This is just the mathematical expression of the operation's linearity.

## 2.4 Measurements

Consider a partition $\{P_\lambda\}$ of the phase space. The functions $P_\lambda(x)$ are indicator-functions over the phase space, taking values 0 or 1. They form a complete set of pairwise disjoint functions:

$$\sum_\lambda P_\lambda \equiv 1, \quad P_\lambda P_\mu = \delta_{\lambda\mu} P_\lambda. \tag{2.9}$$

We consider the indicator functions as binary physical quantities. The whole variety of *physical quantities* is represented by real functions $A(x)$ on the phase space. Each physical quantity $A$ possesses, in arbitrary good approximation, the step-function expansion

$$A(x) = \sum_\lambda A_\lambda P_\lambda(x); \qquad \lambda \neq \mu \Rightarrow A_\lambda \neq A_\mu. \tag{2.10}$$

The real values $A_\lambda$ are step-heights of the function $A(x)$, and $\{P_\lambda\}$ is a partition of the phase space according to them.

The projective partition (2.9) can be generalized. We define a positive decomposition of the constant function:

$$1 = \sum_n \Pi_n(x); \quad \Pi_n(x) \geq 0. \tag{2.11}$$

The elements of the positive decomposition, also called effects, are non-negative functions $\Pi_n(x)$. They need be neither disjoint functions nor indicator-functions at all. They are, in a sense, the unsharp version of indicator-functions.

### *2.4.1 Projective Measurement*

On each system in a statistical ensemble of state $\rho$, we can measure the simultaneous values of the indicator-functions $P_\lambda$ of a given partition (2.9). The outcomes are random. One of the binary quantities, say $P_\lambda$, is 1 with probability

$$p_\lambda = \int P_\lambda \rho \mathrm{d}x, \tag{2.12}$$

while the rest of them is 0:

$$\begin{array}{cccccccc} P_1 & P_2 & \dots & P_{\lambda-1} & P_\lambda & P_{\lambda+1} & \dots \\ \downarrow & \downarrow & & \downarrow & \downarrow & \downarrow & & . \\ 0 & 0 & \dots & 0 & 1 & 0 & \dots \end{array} \tag{2.13}$$

The state suffers projection according to Bayes theorem of conditional probabilities:

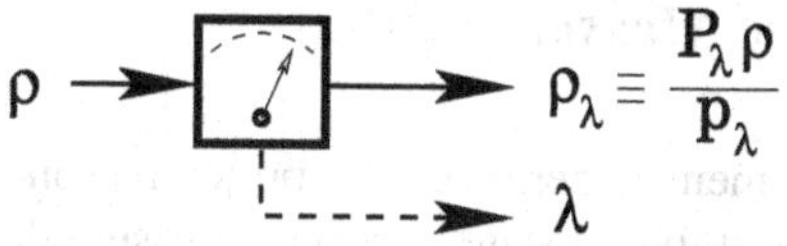


**Fig. 2.5** Selective measurement. The ensemble of pre-measurement states $\rho$ is selected into sub-ensembles of conditional post-measurement states $\rho_\lambda$ according to the obtained measurement outcomes $\lambda$. The probability $p_\lambda$ coincides with the norm of the unnormalized conditional state $P_\lambda \rho$

$$\rho \to \rho_\lambda \equiv \frac{1}{p_\lambda} P_\lambda \rho. \tag{2.14}$$

The post-measurement state $\rho_\lambda$ is also called conditional state, i.e., conditioned on the random outcome $\lambda$. As a result of the above measurement we have randomly selected the original ensemble of state $\rho$ into sub-ensembles of states $\rho_\lambda$ for $\lambda = 1, 2, ....$

The projective measurement is repeatable. Repeated measurements of the indicator functions $P_\mu$ on $\rho_\lambda$ yield always the former outcomes $\delta_{\lambda\mu}$. The above selection is also reversible. If we re-unite the obtained sub-ensembles, the post-measurement state becomes the following mixture of the conditional states $\rho_\lambda$:

$$\sum_\lambda p_\lambda \rho_\lambda = \sum_\lambda p_\lambda \frac{1}{p_\lambda} P_\lambda \rho = \rho. \tag{2.15}$$

This is, of course, identical to the original pre-measurement state.

By the projective measurement of a general physical quantity $A$ we mean the projective measurement of the partition (2.9) generated by its step-function-expansion (2.10). The measured value of $A$ is one of the step-heights:

$$A \to A_\lambda, \tag{2.16}$$

the probability of the particular outcome being given by (2.12). The projective measurement is always repeatable. If a first measurement yielded $A_\lambda$ on a given state then also the repeated measurement yields $A_\lambda$. We can define the non-selective measured value of $A$, i.e., the average of $A_\lambda$ taken with the distribution (2.12):

$$\langle A \rangle \equiv \sum_\lambda p_\lambda A_\lambda = \int A \rho \mathrm{d}x. \tag{2.17}$$

This is also called the *expectation value* of $A$ in the state $\rho$.

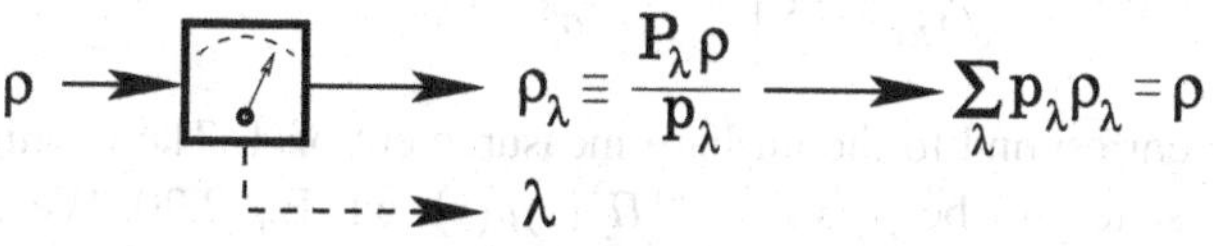


**Fig. 2.6** Non-selective measurement. The sub-ensembles of conditional post-measurement states $\rho_\lambda$ are re-united, contributing to the ensemble of non-selective post-measurement state which is, obviously, identical to the pre-measurement state $\rho$

### 2.4.2 Non-Projective Measurement

Non-projective measurement generalizes the projective one Sect 2.4.1. On each system in a statistical ensemble of state $\rho$, we can measure the simultaneous values of the effects $\Pi_n$ of a given positive decomposition (2.11) but we lose repeatability of the measurement. The outcomes are random. One of the effects, say $\Pi_n$, is 1 with probability

$$p_n = \int \Pi_n \rho \mathrm{d}x, \tag{2.18}$$

while the rest of them is 0:

$$\begin{array}{cccccccc} \Pi_1 & \Pi_2 & \ldots & \Pi_{n-1} & \Pi_n & \Pi_{n+1} & \ldots \\ \downarrow & \downarrow & & \downarrow & \downarrow & \downarrow & \\ 0 & 0 & \ldots & 0 & 1 & 0 & \ldots \end{array} \tag{2.19}$$

The state suffers a change according to the Bayes theorem of conditional probabilities:

$$\rho \to \rho_n \equiv \frac{1}{p_n} \Pi_n \rho. \tag{2.20}$$

Contrary to the projective measurements, the repeated non-projective measurements yield different outcomes in general. The effects $\Pi_n$ are not binary quantities. The individual measurement outcomes 0 or 1 provide unsharp information that can only orient the outcome of subsequent measurements. Still, the selective non-projective measurements are reversible. Re-uniting the obtained sub-ensembles, i.e., averaging the post-measurement conditional states $\rho_n$, yields the original pre-measurement state.

### 2.4.3 Weak Measurement, Time-Continuous Measurement

We can easily generalize the discrete set of effects to continuous sets. This generalization has a merit: one can construct the unsharp measurement of an arbitrarily chosen physical quantity $A$. One constructs the following set of effects:

$$\Pi_{\bar{A}}(x) = \frac{1}{\sqrt{2\pi\sigma^2}} \exp\left[-\frac{(\bar{A} - A(x))^2}{2\sigma^2}\right], \quad -\infty \leq \bar{A} \leq \infty. \tag{2.21}$$

These effects correspond to the unsharp measurement of $A$. The conditional post-measurement state will be $\rho_{\bar{A}}(x) = p_{\bar{A}}^{-1} \Pi_{\bar{A}}(x) \rho(x)$, c.f. Eq. 2.20. We interprete $\bar{A}$ as the random outcome representing the measured value of $A$ at the standard measurement error $\sigma$. The outcome probability (2.18) turns out to be the following distribution function:

$$p_{\bar{A}} = \int \Pi_{\bar{A}}(x)\rho(x)\mathrm{d}x, \tag{2.22}$$

normalized obviously by $\int p_{\bar{A}}\mathrm{d}\bar{A} = 1$.

The behaviour of the unsharp measurement simplifies in the weak measurement limit $\sigma\to\infty$ [1]. This limit means in practice that the accuracy $\sigma$ must be chosen much poorer than the maximum stochastic spread of the measured quantity $A$ in the given state $\rho$. Then the distribution (2.22) can be approximated by a Gaussian centered at $\langle A\rangle$:

$$p_{\bar{A}} \approx \frac{1}{\sqrt{2\pi\sigma^2}}\exp\left[-\frac{(\bar{A}-\langle A\rangle)^2}{2\sigma^2}\right]. \tag{2.23}$$

A useful simplification has been achieved at the price that the precision of a single weak measurement is extremely poor. This incapacity can be compensated fully by a suitable large statistics of repeated weak measurements.

This is the case, e.g., in time-continuous measurement of a given quantity $A$, performed by monitoring a *single* system. Intuitively, we can consider measurements of $\hat{A}$ repeated at frequency $1/\Delta t$ and then we might take the infinite frequency limit $\Delta t\to 0$ of very unsharp - weak - measurements. The error $\sigma$ of single weak measurements must be proportional to their frequency $1/\Delta t$ of repetition. The rate

$$g = \lim_{\Delta t\to 0,\sigma\to\infty} \frac{1}{\Delta t\sigma^2} \tag{2.24}$$

is called the strength of the continuous measurement. It is known that such a construction of time-continuous measurement does really work [2]. For completeness, we include the resulting stochastic equations

$$\frac{\mathrm{d}}{\mathrm{d}t}\rho = \{H,\rho\} - \sqrt{g}w(A-\langle A\rangle)\rho, \tag{2.25}$$

$$\bar{A} = \langle A\rangle + \frac{w}{\sqrt{g}}, \tag{2.26}$$

$w$ is the standard white-noise function: $\langle w(t)\rangle = 0$ and $\langle w(t)w(s)\rangle = \delta(t-s)$. The Eq. 2.25 governs the evolution of the state $\rho$ under monitoring the quantity $A$; the Eq. 2.26 shows that the time-dependent outcome (measurement signal) $\bar{A}$ is always centered at the expectation value in the current state apart from a white-noise whose intensity is inversely proportional to the measurement strength $g$.

## 2.5 Composite Systems

The phase space of the composite system, composed of the subsystems $A$ and $B$, is the Cartesian product of the phase spaces of the subsystems:

$$\Gamma_{AB} = \Gamma_A \times \Gamma_B = \{(x_A, x_B)\}. \tag{2.27}$$

The state of the composite system is described by the normalized distribution function depending on both phase points $x_A$ and $x_B$:

$$\rho_{AB} = \rho_{AB}(x_A, x_B). \tag{2.28}$$

The reduced state of subsystem $A$ is obtained by integration of the composite system's state over the phase space of the subsystem $B$:

$$\rho_A = \int \rho_{AB} \mathrm{d}x_B \equiv \mathcal{M}\rho_{AB}. \tag{2.29}$$

Our notation indicates that reduction, too, can be considered as an operation: it maps the states of the original system into the states of the subsystem. The state $\rho_{AB}$ of the composite system is the product of the subsystem's states if and only if there is no statistical correlation between the subsystems. But generally there is some:

$$\rho_{AB} = \rho_A \rho_B + \text{cl. corr.} \tag{2.30}$$

Nevertheless, the state of the composite system is always *separable*, i.e., we can prepare it as the statistical mixture of product (uncorrelated) states:

$$\rho_{AB}(x_A, x_B) = \sum_\lambda w_\lambda \rho_{A\lambda}(x_A) \rho_{B\lambda}(x_B), \quad w_\lambda \geq 0, \sum_\lambda w_\lambda = 1. \tag{2.31}$$

The equation of motion of the composite system reads

$$\frac{\mathrm{d}}{\mathrm{d}t}\rho_{AB} = \{H_{AB}, \rho_{AB}\}. \tag{2.32}$$

The composite Hamilton function is the sum of the Hamilton functions of the subsystems themselves plus the interaction Hamilton function:

$$H_{AB}(x_A, x_B) = H_A(x_A) + H_B(x_B) + H_{ABint}(x_A, x_B). \tag{2.33}$$

If $H_{ABint}$ is zero then the product initial state remains product state, the dynamics does not create correlation between the subsystems. Non-vanishing $H_{ABint}$ does usually create correlation. The motion of the whole system is reversible, of course. But that of the subsystems is not. In case of product initial state $\rho_A(0)\rho_B(0)$, for instance, the *reduced dynamics* of the subsystem $A$ will represent the time-dependent irreversible[2] operation $\mathcal{M}_A(t)$ which we can formally write as

[2] Note that here and henceforth we use the notion of irreversibility as an equivalent to non-invertibility. We discuss the entropic-informatic notion of irreversibility in Sect. 9.8.

$$\rho_A(t) = \int \rho_A(0)\rho_B(0) \circ U_{AB}^{-1}(t)\mathrm{d}x_B \equiv \mathcal{M}_A(t)\rho_A(0). \tag{2.34}$$

The reversibility of the composite state dynamics has become lost by the reduction: the final reduced state $\rho_A(t)$ does not determine a unique initial state $\rho_A(0)$.

## 2.6 Collective System

The state (2.2) of a system is interpreted on the statistical ensemble of identical systems in the same state. We can form a multiple composite system from a big number $n$ of such identical systems. This we call collective system; its state space is the $n$-fold Cartesian product of the elementary subsystems phase spaces:

$$\Gamma \times \Gamma \times \ldots \Gamma \equiv \Gamma^{\times n}. \tag{2.35}$$

The collective state reads

$$\rho(x_1)\rho(x_2)\ldots\rho(x_n) \equiv \rho^{\times n}(x_1, x_2, \ldots, x_n). \tag{2.36}$$

If $A(x)$ is a physical quantity of the elementary subsystem then, in a natural way, one can introduce its arithmetic mean, over the $n$ subsystems, as a collective physical quantity

$$\frac{A(x_1) + A(x_2) + \ldots + A(x_n)}{n}. \tag{2.37}$$

Collective physical quantities are not necessarily of such simple form. Their measurement is the collective measurement. It can be reduced to independent measurements on the $n$ subsystems.

## 2.7 Two-State System (Bit)

Consider a system of a single degree of freedom, possessing the following Hamilton function:

$$H(q,p) = \frac{1}{2}p^2 + \frac{\omega^2}{8a^2}\left(q^2 - a^2\right)^2. \tag{2.38}$$

The "double-well" potential has two symmetric minima at places $q = \pm a$, and a potential barrier between them. If the energy of the system is smaller than the barrier then the system is localized in one or the other well, moving there periodically "from wall to wall". If, what is more, the energy is much smaller than the barrier height then the motion is restricted to the narrow parts around $q = a$ or $q = -a$, respectively, whereas the motion "from wall to wall" persists always. In that restricted sense has the system two-states.

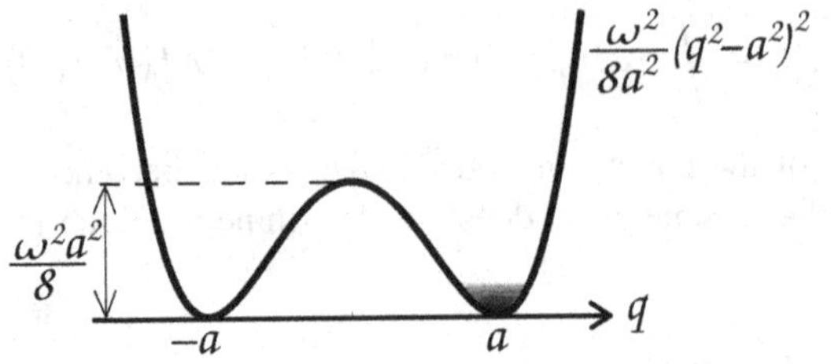


**Fig. 2.7** Classical "two-state system" in double-well potential. The picture visualizes the state concentrated in the r.h.s. well. It is a mixture of periodic "from wall to wall" orbits of various energies that are still much smaller than the barrier height $\omega^2 a^2/8$. One can simplify this low energy regime into a discrete two-state system without the dynamics. The state space becomes discrete consisting of two points associated with $\mathrm{x} = 0$ and $\mathrm{x} = 1$ to store physically what will be called a bit x

One unit of information, i.e. one *bit*, can be stored in it. The localized motional state around $q = -a$ can be associated with the value 0 of a binary digit x, while that around $q = a$ can be associated with the value 1. The information storage is still perfectly reliable if we replace pure localized states and use their mixtures instead. However, the system is more protected against external perturbations if the localized states constituting the mixture are all much lower than the barrier height.

The original continuous phase space (2.1) of the system has thus been restricted to the discrete set $\mathrm{x} = \{0,1\}$ of two elements. Also the states (2.2) have become described by the discrete distribution $\rho(\mathrm{x})$ normalized as $\sum_{\mathrm{x}}\rho(\mathrm{x}) = 1$. There are only two pure states (2.3), namely $\delta_{\mathrm{x}0}$ or $\delta_{\mathrm{x}1}$. To treat classical information, the concept of *discrete* state space will be essential in Chap. 9. In the general case, we use states $\rho(\mathrm{x})$ where x is an integer of, say, $n$ binary digits. The corresponding system is a composite system of $n$ bits.

## 2.8 Problems, Exercises

2.1 *Mixture of pure states.* Let $\rho$ be a mixed state which we mix from pure states. What are the weights we must take for the pure states, respectively? Let us start the solution with the two-state system.

2.2 *Probabilistic or deterministic mixing?* What happens if the mixing is not randomly performed? Let the target state of mixing be evenly distributed: $\rho(\mathrm{x}) = 1/2$. Let someone mix an *equal* number $n$ of the pure states $\delta_{\mathrm{x}0}$ and $\delta_{\mathrm{x}1}$, respectively. Let us write down the state of this $n$–fold composite system. Let us compare it with the $n$–fold composite state corresponding to the proper, i.e. random, mixing.

2.3 *Classical separability.* Let us prove that a classical composite system is always separable. Method: let the index $\lambda$ in (2.31) run over the phase space (2.1) of the composite system. Let us choose $\lambda = (\bar{x}_A, \bar{x}_B)$.

2.4 *Decorrelating a single state?* Does an operation $\mathcal{M}$ exist such that it brings an arbitrary correlated state $\rho_{AB}$ into the (uncorrelated) product state $\rho_A\rho_B$ of the reduced states $\rho_A$ and $\rho_B$? Remember, the operation $\mathcal{M}$ must be linear.

2.5 *Decorrelating an ensemble.* Give an operation $\mathcal{M}$ that brings $2n$ correlated states $\rho_{AB}$ into $n$ uncorrelated states $\rho_A\rho_B$: $\mathcal{M}\rho_{AB}^{\times 2n} = (\rho_A\rho_B)^{\times n}$. Method: consider a smart permutation of the $2n$ copies of the subsystem $A$, followed by a reduction to the suitable subsystem.

2.6 *Measurement and Bayes theorem.* The heart of the classical theory of estimation is the Bayes theorem. Let us prove that the measurement scheme is equivalent to it.

2.7 *Indirect measurement.* Let us prove that the non-projective measurement of arbitrarily given effects $\{\Pi_n(x)\}$ can be obtained from projective measurements on a suitably enlarged composite state. Method: Construct the suitable composite state $\rho(x, n)$ to include a hypothetical detector system to count $n$; perform projective measurement on the detector's $n$.

## References

1. Aharonov, Y., Albert, D.Z., Vaidman, L.: Phys. Rev. Lett. **60**,1351 (2008)
2. Diósi, L.: Weak measurements in quantum mechanics. In: Françoise, J.P., Naber, G.L., Tso, S.T. (eds) Encyclopedia of Mathematical Physics, vol. 4, pp. 276–282. Elsevier, Oxford (2006)

# Chapter 3
# Semiclassical, Semi-Q-Physics

The dynamical laws of classical physics, given in Chap. 2, can approximatively be retained for microscopic systems as well, but with restrictions of a new type. The basic goal is to impose discreteness onto the classical theory. We add discretization q-conditions to the otherwise unchanged classical canonical equations. The corresponding restrictions must be graceful in the sense that they must not modify the dynamics of macroscopic systems and they must not destroy the consistency of the classical equations.

Let us assume that the dynamics of the microsystem is separable in the canonical variables $(q_k, p_k)$, and the motion is finite in phase space. The canonical action variables are defined as

$$I_k \equiv \frac{1}{2\pi}\oint p_k \mathrm{d}q_k, \tag{3.1}$$

for all degrees of freedom $k = 1, 2, \ldots$ The integral is understood along one period of the finite motion in each degree of freedom. The action variables $I_k$ are the adiabatic invariants[1] of classical motion. In classical physics they can take arbitrary values. To impose discreteness on classical dynamics, the Bohr–Sommerfeld q-condition says that each action $I_k$ must be an integer multiple of the Planck constant (plus $\hbar/2$ in case of oscillatory motion):

$$I_k \equiv \frac{1}{2\pi}\oint p_k \mathrm{d}q_k = \left(n_k + \frac{1}{2}\right)\hbar. \tag{3.2}$$

The integer q-numbers $n_k$ will label the discrete sequence of phase space trajectories which are, according to this semiclassical theory, the only possible motions. The state with $n_1 = n_2 = \ldots = 0$ is the ground state and the excited states are separated by finite energy gaps from it.

[1] See, e.g., in Chap. VII. of Landau and Lifshitz [1].

L. Diósi, *A Short Course in Quantum Information Theory*,
Lecture Notes in Physics, 827, DOI: 10.1007/978-3-642-16117-9_3,

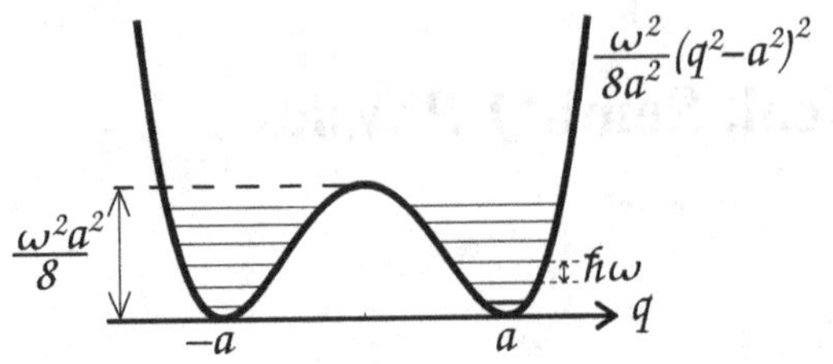


**Fig. 3.1** Stationary q-states in double-well potential. The bottoms of the wells can be approximated by quadratic potentials $\frac{1}{2}\omega^2(q \mp a)^2$. Thus we obtain the energy-level structure of two separate harmonic oscillators, one in the l.h.s. well, the other in the r.h.s. well. This approximation breaks down for the upper part of the wells. Perfect two-state q-systems will be realized at low energies where the degenerate ground states never get excited

Let us consider the double-well potential (2.38) with suitable parameters such that the lowest states be doubly degenerate, of approximate energies $\hbar\omega, 2\hbar\omega, 3\hbar\omega$ etc., localized in either the left- or the right-side well. The parametric condition is that the barrier be much higher than the energy gap $\hbar\omega$.

Let us store 1 bit of information in the two ground states, say the ground state in the left-side well means 0 and that in the right-side means 1. These two states are separated from all other states by a minimum energy $\hbar\omega$. Perturbations of energies smaller than $\hbar\omega$ are not able to excite the two ground states. In this sense the above system is a perfect autonomous two-state system provided the energy of its environment is sufficiently low. This autonomy follows from quantization and is the property of q-systems.

The Bohr–Sommerfeld theory classifies the possible stationary states of dynamically separable microsystems.[2] It remains in debt of capturing non-stationary phenomena. The true q-theory ( Chap. 4) will come to the decision that the generic, non-stationary, states emerge from superposition of the stationary states. In case of the above two-state system, the two ground states must be considered as the two orthonormal vectors of a two-dimensional complex vector space. Their normalized complex linear combinations will represent all states of the two-state quantum system. This q-system and its continuum number of states will constitute the ultimate notion of q-bit or *qubit*.

## 3.1 Problems, Exercises

3.1 *Bohr quantization of the harmonic oscillator.* Let us derive the Bohr–Sommerfeld q-condition for the one-dimensional harmonic oscillator of mass $m = 1$, bounded by the potential $\frac{1}{2}\omega^2 q^2$.

---

[2] The modern semiclassical theory is more general and powerful, cf. Gutzwiller [2].

3.2 *The role of adiabatic invariants.* Consider the motion of the harmonic oscillator that satisfies the q-conditions with a certain q-number $n$. Suppose that we are varying the directional force constant $\omega^2$ adiabatically, i.e., much slower than one period of oscillation. Physical intuition says that the motion of the system should invariably satisfy the q-condition to good approximation, even with the same q-number $n$. Is that true?

3.3 *Classical-like or q-like motion.* There is no absolute rule to distinguish between microscopic and macroscopic systems. It makes more sense to ask if a given state (motion) is q-like or classical-like. In semiclassical physics, the state is q-like if the q-condition imposes physically relevant restrictions, and the state is classical-like if the imposed discreteness does not practically restrict the continuum of classical states. Let us argue that, in this sense, small integer q-numbers $n$ mean q-like states and large ones mean classical-like states.

## References

1. Landau, L.D., Lifshitz, E.M.: Course of Theoretical Physics: Mechanics. Pergamon, Oxford (1960)
2. Gutzwiller, M.C.: Chaos in Classical and Quantum Mechanics. Springer, Berlin (1991)

# Chapter 4
# Foundations of Q-Physics

We present the standard q-theory [1] while, at each element, striving for the maximum likeness to Chap. 2 on foundations of classical physics. We go slightly beyond the traditional treatment and, e.g., we define *non-projective* q-measurements as well as the phenomenon of *entanglement*. Leaf through Chap. 2 again, and compare!

## 4.1 State Space, Superposition, Equation of Motion

The state space of a q-system is a Hilbert space $\mathcal{H}$. In the case of $d$-state q-system it is the $d$-dimensional complex vector space

$$\mathcal{H} = C^d = \{c_\lambda; \lambda = 1, 2, \ldots, d\}, \tag{4.1}$$

where the $c_k$'s are the elements of the complex column-vector in the given basis. The pure state of a q-system is described by a complex unit vector, also called state vector. In basis-independent abstract (Dirac-) notation it reads

$$|\psi\rangle \equiv \begin{bmatrix} c_1 \\ c_2 \\ \cdot \\ \cdot \\ c_d \end{bmatrix}, \quad \langle\psi| \equiv \left[c_1^*, c_2^*, \ldots, c_d^*\right], \quad \sum_{\lambda=1}^{d} |c_\lambda|^2 = 1. \tag{4.2}$$

The inner product of two vectors is denoted by $\langle\psi|\varphi\rangle$. Matrices are denoted by a "hat" over the symbols, and their matrix elements are written as $\langle\psi|\hat{A}|\varphi\rangle$. In q-theory, the components $c_k$ of the complex vector are called probability amplitudes. Superposition, i.e. normalized complex linear combination of two or more vectors, yields again a possible pure state.

L. Diósi, *A Short Course in Quantum Information Theory*, Lecture Notes in Physics, 827, DOI: 10.1007/978-3-642-16117-9_4,

$\hat{\rho}$ 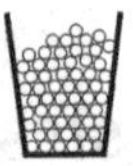

**Fig. 4.1** Urn model of q-statistical ensemble. To visualize the statistical ensemble for a q-state $\hat{\rho}$, we imagine an urn to contain a very large number of copies of the given q-system. The preparation of such ensemble is never unique if the state is mixed

The generic state is mixed, described by trace-one positive semidefinite density matrix

$$\hat{\rho} = \{\rho_{\lambda\mu}; \lambda, \mu = 1, 2, \ldots, d\} \geq 0, \quad \mathrm{tr}\,\hat{\rho} = 1. \tag{4.3}$$

The generic state is interpreted on the statistical ensemble of identical systems. The density matrix of the pure state (4.2) is a special case, it is the one-dimensional hermitian projector onto the subspace given by the state vector:

$$\hat{\rho}_{\mathrm{pure}} = \hat{P} = |\psi\rangle\langle\psi|. \tag{4.4}$$

We can see that multiplying the state vector by a complex phase factor yields the same density matrix, i.e., the same q-state. Hence the phase of the state vector can be deliberately altered, still the same pure q-state is obtained. In the conservative q-theory, contrary to the classical theory, not even the pure state is interpreted on a single system but on the statistical ensemble of identical systems.

Dynamical evolution[1] of a closed q-system is determined by its hermitian Hamilton matrix $\hat{H}$. The von Neumann equation of motion takes this form:

$$\frac{\mathrm{d}\hat{\rho}}{\mathrm{d}t} = -\frac{i}{\hbar}[\hat{H}, \hat{\rho}]. \tag{4.5}$$

For pure states, this is equivalent to the Schrödinger equation of motion

$$\frac{\mathrm{d}|\psi\rangle}{\mathrm{d}t} = -\frac{\mathrm{i}}{\hbar}\hat{H}|\psi\rangle. \tag{4.6}$$

Its solution $|\psi(t)\rangle \equiv \hat{U}(t)|\psi(0)\rangle$ represents a time-dependent unitary map $\hat{U}(t)$ of the Hilbert space. It enables us to construct the solution of the von Neumann equation (4.5):

$$\hat{\rho}(t) = \hat{U}(t)\hat{\rho}(0)\hat{U}^{\dagger}(t) \equiv \mathcal{M}(t)\hat{\rho}(0), \tag{4.7}$$

where $\mathcal{M}(t)$ denotes the corresponding reversible q-operation. Below is introduced the generic notion of q-operation.

[1] Our lectures use the Schrödinger-picture: the q-states $\hat{\rho}$ evolve with $t$, the q-physical quantities $\hat{A}$ do not.

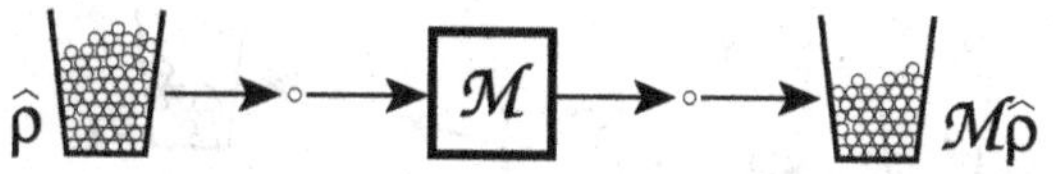


**Fig. 4.2** Urn model of q-operation. We draw a q-system of state $\hat{\rho}$ from the first urn, let it go through the transformator $\mathcal{M}$ and place it into the second urn. We repeat this procedure to reach a very large number of copies in the second urn which will then represent the transformed q-state $\mathcal{M}\hat{\rho}$

## 4.2 Operation, Mixing, Selection

Let operation $\mathcal{M}$ on a given q-state $\hat{\rho}$ mean that we perform the same transformation on each q-system of the corresponding statistical ensemble. Mathematically, $\mathcal{M}$ is linear trace-preserving completely positive map, cf. Sect. 8.1, transforming an arbitrary state $\hat{\rho}$ into a new state $\mathcal{M}\hat{\rho}$. Contrary to classical operations, not all positive maps correspond to realizable q-operations, but the completely positive ones do Fig 4.2.

Randomly mixing the elements of two ensembles of q-states $\hat{\rho}_1$ and $\hat{\rho}_2$ at respective rates $w_1 \geq 0$ and $w_2 \geq 0$ yields the new ensemble of the q-state

$$\hat{\rho} = w_1\hat{\rho}_1 + w_2\hat{\rho}_2; \quad w_1 + w_2 = 1. \tag{4.8}$$

A generic mixed q-state can always be prepared (i.e. decomposed) as the mixture of two or more other mixed q-states in infinitely many different ways. After mixing, however, it is totally impossible to distinguish in which way the mixed q-state was prepared. It is crucial, of course, that mixing must be probabilistic. A given mixed q-state can also be prepared (decomposed) as a mixture of pure q-states and this mixture is, contrary to the classical case, not unique in general.

Selection of a given ensemble into specific sub-ensembles, a contrary process to mixing, will be possible via so-called selective q-operations. They correspond mathematically to trace-reducing completely positive maps, cf. Sect. 8.3. The most typical selective q-operations are called q-measurements, cf. Sect. 4.4.

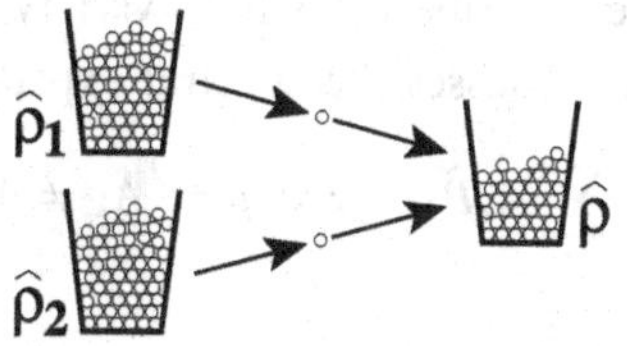


**Fig. 4.3** Urn model of mixing. We draw a q-system of state $\hat{\rho}_1$ or $\rho_2$ from the two urns on the lhs chosen randomly at respective probabilities $w_1$ and $w_2$; we place it into the urn on the rhs. We repeat this procedure to reach a very large number of copies in the urn on the rhs, which will then represent the mixed q-state $\hat{\rho} = w_1\hat{\rho}_1 + w_2\hat{\rho}_2$

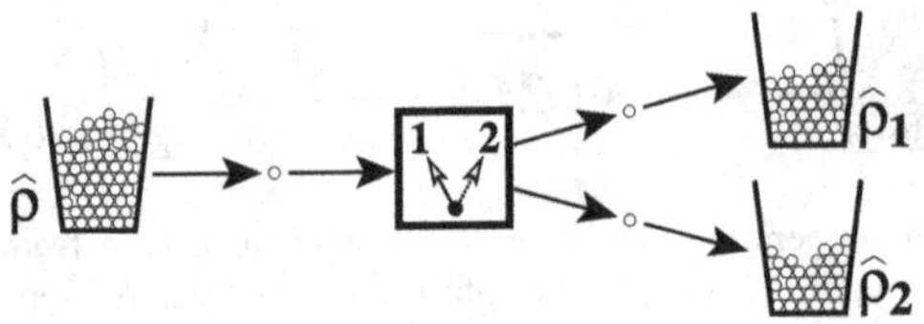


**Fig. 4.4** Urn model of selection. We draw a q-system of state $\hat{\rho}$ from the urns on the lhs and let it go through the selective q-operation, e.g., a measurement apparatus; we place the system into the first or second urns on the rhs according to the outcome 1 or 2 of the measurement. We repeat this procedure to reach a very large number of copies in the urns on the rhs, which will then represent the selected q-states $\hat{\rho}_1$ and $\hat{\rho}_2$, respectively

## 4.3 Linearity of Non-Selective Operations

The q-operation's categorical linearity follows from the linearity of the procedure of mixing (4.8). Obviously we must arrive at the same q-state if we mix two states first and then we subject the systems of the resulting ensemble to the operation $\mathcal{M}$ or, alternatively, we perform the operation prior to mixing the two ensembles together:

$$\mathcal{M}(w_1\hat{\rho}_1 + w_2\hat{\rho}_2) = w_1\mathcal{M}\hat{\rho}_1 + w_2\mathcal{M}\hat{\rho}_2. \tag{4.9}$$

This is just the mathematical expression of the operation's linearity.

## 4.4 Measurements

Consider a partition $\{\hat{P}_\lambda\}$ of the state space. The hermitian matrices $\hat{P}_\lambda$ are projectors on the Hilbert space, of eigenvalues 0 and 1. They form a complete orthogonal set:

$$\sum_\lambda \hat{P}_\lambda = \hat{I}, \quad \hat{P}_\lambda \hat{P}_\mu = \delta_{\lambda\mu}\hat{P}_\lambda. \tag{4.10}$$

We consider the projectors as binary q-physical quantities. Hermitian matrices $\hat{A}$ acting on the Hilbert space describe the whole variety of q-physical quantities. Each q-physical quantity $\hat{A}$ possesses the spectral expansion[2]

$$\hat{A} = \sum_\lambda A_\lambda \hat{P}_\lambda; \quad \lambda \neq \mu \Rightarrow A_\lambda \neq A_\mu. \tag{4.11}$$

[2] Equivalent terminologies, like spectral or diagonal decomposition, or just diagonalization, are in widespread use.

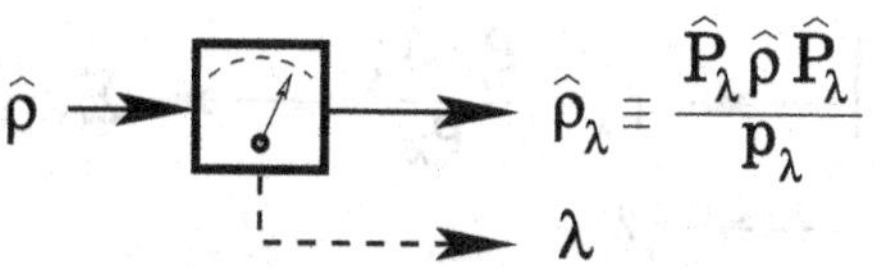


**Fig. 4.5** Selective q-measurement. The ensemble of pre-measurement q-states $\hat{\rho}$ is selected into sub-ensembles of conditional post-measurement q-states $\hat{\rho}_\lambda$ according to the obtained measurement outcomes $\lambda$. The probability $p_\lambda$ coincides with the trace of the unnormalized conditional q-state $\hat{P}_\lambda\hat{\rho}\hat{P}_\lambda$. The relative phases of the sub-ensembles $\hat{\rho}_\lambda$ have been irretrievably lost in a mechanism called decoherence

The real values $A_\lambda$ are eigenvalues of the matrix $\hat{A}$, and $\{\hat{P}_\lambda\}$ is a partition of the state space according to them.

The projective partition (4.10) can be generalized. We define a positive decomposition of the unit matrix:

$$\hat{I} = \sum_n \hat{\Pi}_n; \quad \hat{\Pi}_n \geq 0. \tag{4.12}$$

The elements of the positive decomposition, also called q-effects, are non-negative matrices $\hat{\Pi}_n$. They need be neither orthogonal nor projectors at all. They are, in a sense, the unsharp version of projectors.

### 4.4.1 *Projective Measurement*

On each q-system in a statistical ensemble of q-state $\hat{\rho}$, we can measure the simultaneous values of the orthogonal projectors $\hat{P}_\lambda$ of a given partition (4.10). The outcomes are random. One of the binary quantities, say $\hat{P}_\lambda$, is 1 with probability

$$p_\lambda = \mathrm{tr}\left(\hat{P}_\lambda\hat{\rho}\right), \tag{4.13}$$

while the rest of them is 0:

$$\begin{array}{cccccccc} \hat{P}_1 & \hat{P}_2 & \ldots & \hat{P}_{\lambda-1} & \hat{P}_\lambda & \hat{P}_{\lambda+1} & \ldots \\ \downarrow & \downarrow & & \downarrow & \downarrow & \downarrow & & . \\ 0 & 0 & \ldots & 0 & 1 & 0 & \ldots \end{array} \tag{4.14}$$

The state suffers projection according to the projection postulate of von Neumann and Lüders:

$$\hat{\rho} \to \hat{\rho}_\lambda \equiv \frac{1}{p_\lambda}\hat{P}_\lambda\hat{\rho}\hat{P}_\lambda. \tag{4.15}$$

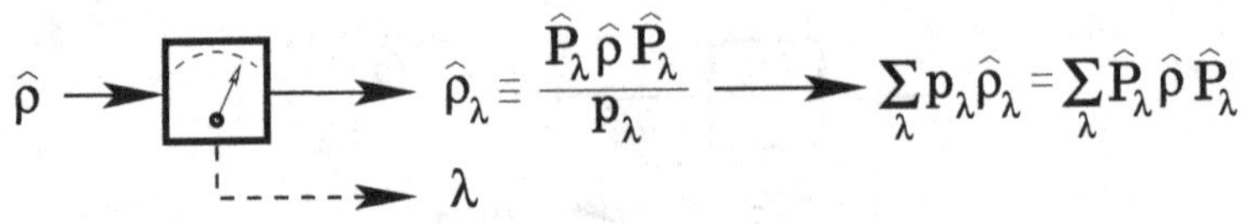


**Fig. 4.6** Non-selective q-measurement. The sub-ensembles of conditional post-measurement q-states $\hat{\rho}_\lambda$ are re-united, contributing to the ensemble of non-selective post-measurement q-state which is, contrary to the classical case and because of decoherence, irreversibly different from the pre-measurement q-state $\hat{\rho}$.

The post-measurement q-state $\hat{\rho}_\lambda$ is also called conditional q-state, i.e., conditioned on the random outcome $\lambda$. As a result of the above measurement we have randomly selected the original ensemble of q-state $\hat{\rho}$ into sub-ensembles of q-states $\hat{\rho}_\lambda$ for $\lambda = 1, 2, \ldots$.

The projective q-measurement is repeatable. Repeated q-measurements of the orthogonal projectors $\hat{P}_\mu$ on $\hat{\rho}_\lambda$ yield always the former outcomes $\delta_{\lambda\mu}$. The above selection is, contrary to the classical case, not reversible. Let us re-unite the obtained sub-ensembles; the post-measurement q-state becomes the following mixture of the conditional q-states $\hat{\rho}_\lambda$ :

$$\sum_\lambda p_\lambda \hat{\rho}_\lambda = \sum_\lambda p_\lambda \frac{1}{p_\lambda} \hat{P}_\lambda \hat{\rho} \hat{P}_\lambda = \sum_\lambda \hat{P}_\lambda \hat{\rho} \hat{P}_\lambda; \quad \neq \hat{\rho}, \tag{4.16}$$

which is in general not identical to the original pre-measurement q-state. The non-selective measurement realizes an irreversible[3] q-operation $\mathcal{M}$:

$$\hat{\rho} \to \sum_\lambda \hat{P}_\lambda \hat{\rho} \hat{P}_\lambda \equiv \mathcal{M}\hat{\rho}. \tag{4.17}$$

Q-irreversibility has its roots in the mechanism of *decoherence*[4] which means the general phenomenon when superpositions get destroyed by q-measurement or by environmental interaction, cf. Chap. 8.

By the projective q-measurement of a general q-physical quantity $\hat{A}$ we mean the projective q-measurement of the partition (4.10) generated by its spectral expansion (4.11). The measured value of $\hat{A}$ is one of the eigenvalues:

$$\hat{A} \to A_\lambda, \tag{4.18}$$

the probability of the particular outcome is given by (4.13). The projective q-measurement is always repeatable. If a first measurement yielded $A_\lambda$ on a given state then also the repeated measurement yields $A_\lambda$. We can define the

---

[3] We use the notion of irreversibility as an equivalent to non-invertibility. We discuss the entropic-informatic notion of q-irreversibility in Sect. 10.8.

[4] Cf., e.g., the monograph by Joos et al. [2].

non-selective measured value of $\hat{A}$, i.e., the average of $A_\lambda$ taken with the distribution (4.13):

$$\langle\hat{A}\rangle \equiv \sum_\lambda p_\lambda A_\lambda = \mathrm{tr}\big(\hat{A}\hat{\rho}\big). \tag{4.19}$$

This is also called the expectation value of $\hat{A}$ in the state $\hat{\rho}$.

### 4.4.2 Non-Projective Measurement

Non-projective q-measurement generalizes the projective one 4.4.1. On each q-system in a statistical ensemble of state $\hat{\rho}$, we can measure the simultaneous values of the q-effects $\hat{\Pi}_n$ of a given positive decomposition (4.12) but we lose repeatability of the measurement. The outcomes are random. One of the q-effects, say $\hat{\Pi}_n$, is 1 with probability

$$p_n = \mathrm{tr}\big(\hat{\Pi}_n\hat{\rho}\big), \tag{4.20}$$

while the rest of them is 0:

$$\begin{array}{ccccccc} \hat{\Pi}_1 & \hat{\Pi}_2 & \dots & \hat{\Pi}_{n-1} & \hat{\Pi}_n & \hat{\Pi}_{n+1} & \dots \\ \downarrow & \downarrow & & \downarrow & \downarrow & \downarrow & \\ 0 & 0 & \dots & 0 & 1 & 0 & \dots \end{array} \tag{4.21}$$

The state suffers the following change[5]:

$$\hat{\rho} \rightarrow \hat{\rho}_n \equiv \frac{1}{p_n}\hat{\Pi}_n^{1/2}\hat{\rho}\hat{\Pi}_n^{1/2}. \tag{4.22}$$

Contrary to the projective q-measurements, the repeated non-projective q-measurements yield different outcomes in general. The q-effects $\hat{\Pi}_n$ are not binary quantities. The individual measurement outcomes 0 or 1 provide unsharp information that can only orient the outcome of subsequent measurements. The selective non-projective q-measurements are obviously irreversible: re-uniting the obtained sub-ensembles, i.e., averaging the post-measurement conditional q-states $\hat{\rho}_n$ yields $\sum_n \hat{\Pi}_n^{1/2}\hat{\rho}\hat{\Pi}_n^{1/2}$ which differs from the original pre-measurement q-state $\hat{\rho}$.

[5] Note that, in the q-literature, the post-measurement states are usually specified in a more general form $(1/p_n)\hat{U}_n\hat{\Pi}_n^{1/2}\hat{\rho}\hat{\Pi}_n^{1/2}\hat{U}_n^\dagger$, to include the arbitrary selective post-measurement unitary transformations $\hat{U}_n$.

### 4.4.3 Weak Measurement, Time-Continuous Measurement

We can easily generalize the discrete set of q-effects to continuous sets. This generalization has a merit: one can construct the unsharp measurement of an arbitrarily chosen quantity $\hat{A}$. One constructs the following set of q-effects:

$$\hat{\Pi}_{\bar{A}} = \frac{1}{\sqrt{2\pi\sigma^2}} \exp\left[-\frac{(\bar{A}-\hat{A})^2}{2\sigma^2}\right], \quad -\infty \leq \bar{A} \leq \infty. \tag{4.23}$$

These q-effects correspond to the unsharp q-measurement of $\hat{A}$. The conditional post-measurement q-state will be $\hat{\rho}_{\bar{A}} = p_{\bar{A}}^{-1}\hat{\Pi}_{\bar{A}}^{1/2}\hat{\rho}\hat{\Pi}_{\bar{A}}^{1/2}$, cf. Eq. (4.22). We interprete $\bar{A}$ as the random outcome representing the measured value of $\hat{A}$ at the standard measurement error $\sigma$. The outcome probability (4.20) turns out to be the following distribution function:

$$p_{\bar{A}} = \mathrm{tr}\left(\hat{\Pi}_{\bar{A}}\hat{\rho}\right), \tag{4.24}$$

normalized obviously by $\int p_{\bar{A}} d\bar{A} = 1$.

The behaviour of the unsharp q-measurement simplifies in the weak measurement limit $\sigma \to \infty$ [3] . This limit means in practice that the accuracy $\sigma$ must be chosen much poorer than the maximum stochastic spread of the measured quantity $\hat{A}$ in the given q-state $\hat{\rho}$. Then the distribution (4.23) can be approximated by a Gaussian centered at $\langle\hat{A}\rangle$ :

$$p_{\bar{A}} \approx \frac{1}{\sqrt{2\pi\sigma^2}} \exp\left[-\frac{(\bar{A}-\langle\hat{A}\rangle)^2}{2\sigma^2}\right]. \tag{4.25}$$

A useful simplification has been achieved at the price that the precision of a single weak measurement is extremely poor. This incapacity can be compensated fully by a suitable large statistics of repeated weak measurements.

This is the case, e.g., in time-continuous q-measurement of a given quantity $\hat{A}$, performed by monitoring a *single* q-system. Intuitively, we can consider measurements of $\hat{A}$ repeated at frequency $1/\Delta t$ and then we might take the infinite frequency limit $\Delta t \to 0$ of very unsharp—weak—measurements. The error $\sigma$ of single weak measurements must be proportional to their frequency $1/\Delta t$ of repetition. The rate

$$g = \lim_{\Delta t \to 0, \sigma \to \infty} \frac{1}{\Delta t \sigma^2} \tag{4.26}$$

is called the strength of the continuous measurement. It is known that such a construction of time-continuous measurement does really work [4]. Rather than presenting the equations of the selective time-continuous q-measurement, this time we consider the non-selective case which is much simpler. Under the above mechanism of continuous measurement of $\hat{A}$ the reversible equation of motion (4.5) turns into the following irreversible q-master equation:

$$\frac{\mathrm{d}\hat{\rho}}{\mathrm{d}t} = -\frac{\mathrm{i}}{\hbar}[\hat{H}, \hat{\rho}] - \frac{g}{8}[\hat{A}, [\hat{A}, \hat{\rho}]]. \tag{4.27}$$

The new double-commutator term on the rhs describes the decoherence caused by the continuous q-measurement. This is a special case of the general q-master equations presented in Sect. 8.6.

Let us see how this new term comes about. Note first that, in the asymptotics of the continuous limit (4.26), the error $\sigma$ in the q-effects (4.23) can be expressed through the measurement strength $g$ and the time-step $\Delta t$:

$$\hat{\Pi}_{\bar{A}} = \sqrt{\frac{g\Delta t}{2\pi}} \exp\left[-\frac{g}{2}(\sqrt{\Delta t}\bar{A} - \sqrt{\Delta t}\hat{A})^2\right]. \tag{4.28}$$

Each time the unsharp measurement of $\hat{A}$ happens we can write the non-selective change of the q-state in the following form:

$$\hat{\rho} \to \int \hat{\Pi}_{\bar{A}}^{1/2} \hat{\rho} \hat{\Pi}_{\bar{A}}^{1/2} \mathrm{d}\bar{A}. \tag{4.29}$$

Then we substitute the q-effects (4.28) and change the integration variable from $\bar{A}$ to $a = \sqrt{\Delta t}\bar{A}$, yielding the following expression for the post-measurement state:

$$\sqrt{\frac{g}{2\pi}} \int \exp\left[-\frac{g}{4}(a - \sqrt{\Delta t}\hat{A})^2\right] \hat{\rho} \exp\left[-\frac{g}{4}(a - \sqrt{\Delta t}\hat{A})^2\right] \mathrm{d}a. \tag{4.30}$$

Since we are interested in the continuous limit, we can restrict the above expression for the leading term $\hat{\rho} - (g/8)\left[\hat{A}, \left[\hat{A}, \hat{\rho}\right]\right]\Delta t$ in the small $\Delta t$. Hence, in the continuous limit $\Delta t \to 0$, we arrive at the new double-commutator contribution to $\mathrm{d}\hat{\rho}/\mathrm{d}t$.

### 4.4.4 Compatible Physical Quantities

Let $\hat{A}$ and $\hat{B}$ be two arbitrary q-physical quantities. Consider their spectral expansions (4.11)

$$\hat{A} = \sum_\lambda A_\lambda \hat{P}_\lambda, \quad \hat{B} = \sum_\mu B_\mu \hat{Q}_\mu. \tag{4.31}$$

Let us measure both q-physical quantities, $\hat{A}$ first and then $\hat{B}$, in subsequent projective measurements. Write down the selective change of the q-state (4.16):

$$\hat{\rho} \rightarrow \frac{\hat{P}_\lambda \hat{\rho} \hat{P}_\lambda}{p_\lambda} \rightarrow \frac{\hat{Q}_\mu \hat{P}_\lambda \hat{\rho} \hat{P}_\lambda \hat{Q}_\mu}{p_{\mu\lambda}}, \tag{4.32}$$

where $p_{\mu\lambda}$ is the probability that $\hat{A} \rightarrow A_\lambda$ is the first, then $\hat{B} \rightarrow B_\mu$ is the second measurement outcome. Obviously, had we performed the two measurements in the reversed order, then the distribution of the measurement outcomes would in general be different. This we interpret in such a way that $\hat{A}$ and $\hat{B}$ are not simultaneously measurable. Yet, in some important cases, the measurement outcomes are independent of the order of measurements. The corresponding condition is that the matrices of the two q-physical quantities commute:

$$\left[\hat{P}_\lambda, \hat{Q}_\mu\right] = 0 \Longleftrightarrow \left[\hat{A}, \hat{B}\right] = 0. \tag{4.33}$$

Q-physical quantities with commuting matrices are called compatible. On compatible q-physical quantities one can obtain simultaneous information. Two arbitrarily chosen q-physical quantities are, however, not compatible in general.

The matrices (projectors) of binary q-physical quantities $\hat{P}_\lambda$, contributing to the spectral expansion (4.11) of a given q-physical quantity, will always commute by definition (4.10), they are compatible, hence they can be measured simultaneously. The projective measurement defined in Sect. 4.4.1 is just the simultaneous measurement of all $\hat{P}_\lambda$'s.

Certain incompatible physical quantities can be measured in non-projective measurements. The q-effects of a positive decomposition (4.12) are not compatible in general, their non-negative matrices $\hat{\Pi}_n$ may not commute. Yet their simultaneous non-projective measurement 4.4.2 is possible though in restricted sense compared to projective measurements 4.4.1: we lose repeatability of the measurement.

### 4.4.5 Measurement in Pure State

General rules of q-measurement (4.13–4.18) simplify significantly for a pure state $\hat{\rho} = |\psi\rangle\langle\psi|$. The measurement outcome is still one of the eigenvalues $A_\lambda$. As for the pure state, it jumps into a conditional pure state $|\psi_\lambda\rangle$:

$$\hat{A} \rightarrow A_\lambda, \quad |\psi\rangle \rightarrow |\psi_\lambda\rangle \equiv \frac{1}{\sqrt{p_\lambda}} \hat{P}_\lambda |\psi\rangle, \tag{4.34}$$

while the probability distribution of the measurement outcome reads

$$p_\lambda = \langle\psi|\hat{P}_\lambda|\psi\rangle. \tag{4.35}$$

The description of the process becomes even simpler if the matrix of the measured physical quantity is non-degenerate:

$$\hat{A} = \hat{A}^{\dagger} = \sum_{\lambda=1}^{d} A_{\lambda}\hat{P}_{\lambda} \Longleftrightarrow \hat{P}_{\lambda} = |\varphi_{\lambda}\rangle\langle\varphi_{\lambda}|. \tag{4.36}$$

In this case, the state vector jumps into the unique eigenvector belonging to the measured eigenvalue:

$$\hat{A} \rightarrow A_{\lambda}, \; |\psi\rangle \rightarrow |\varphi_{\lambda}\rangle. \tag{4.37}$$

And the probability distribution of the measurement outcome can be written as the squared modulus of the inner product between the new and the old state vectors:

$$p_{\lambda} = \langle\psi| \cdot |\varphi_{\lambda}\rangle\langle\varphi_{\lambda}| \cdot |\psi\rangle = |\langle\varphi_{\lambda}|\psi\rangle|^2. \tag{4.38}$$

Most often, the description of the quantum measurement happens in such a way that we expand the pre-measurement pure state in terms of the eigenvectors of the physical quantity to be measured:

$$|\psi\rangle = \sum_{\lambda} c_{\lambda}|\varphi_{\lambda}\rangle. \tag{4.39}$$

In this representation the measurement takes place this way:

$$\hat{A} \rightarrow A_{\lambda}, \quad \begin{bmatrix} c_1 \\ \cdot \\ c_{\lambda} \\ \cdot \\ c_d \end{bmatrix} \rightarrow \begin{bmatrix} 0 \\ 0 \\ 1 \\ 0 \\ 0 \end{bmatrix}, \tag{4.40}$$

while the probability distribution of the measurement outcome reads

$$p_{\lambda} = |c_{\lambda}|^2. \tag{4.41}$$

## 4.5 Composite Systems

The state space of the composite q-system, composed of the q-subsystems $A$ and $B$, is the tensor product of the vector spaces of the q-subsystems, cf. (2.27):

$$\mathcal{H}_{AB} = C^{d_A} \otimes C^{d_B} = \{c_{\lambda l}; \lambda = 1, \ldots, d_A; l = 1, \ldots, d_B\}. \tag{4.42}$$

The pure state of the composite system is described by the normalized state vector of dimension $d_A d_B$:

$$|\psi_{AB}\rangle = \{c_{\lambda l}\}; \quad \sum_{\lambda,l} |c_{\lambda l}|^2 = 1, \tag{4.43}$$

If the two subsystems are uncorrelated then the state vector of the composite system is a tensor product

$$|\psi_{AB}\rangle = |\psi_A\rangle \otimes |\psi_B = \{c_{A\lambda} c_{Bl}\}. \tag{4.44}$$

The mixed state of the composite system is described by the $d_A d_B \times d_A d_B$-dimensional density matrix (4.3)

$$\hat{\rho}_{AB} = \{\rho_{(\lambda l)(\mu m)}\}. \tag{4.45}$$

The reduced state of subsystem $A$ is obtained by tracing the composite q-system's state over the Hilbert space of subsystem $B$:

$$\hat{\rho}_A = \mathrm{tr}_B \hat{\rho}_{AB} = \left\{ \sum_l \rho_{(\lambda l)(\mu l)} \right\} \equiv \mathcal{M} \hat{\rho}_{AB}. \tag{4.46}$$

Our notation indicates that a reduction, too, can be considered as an operation $\mathcal{M}$: it maps the states of the original q-system into the states of the q-subsystem. The state $\hat{\rho}_{AB}$ of the composite q-system is the tensor product of the q-subsystem's states if and only if there is no statistical correlation between the subsystems. But generally there is some, and then there can be *q-correlation* as well. Symbolically, we write

$$\hat{\rho}_{AB} = \hat{\rho}_A \otimes \hat{\rho}_B + \text{cl. corr.} + \textbf{q-corr.}, \tag{4.47}$$

which is different from the classical correlations (2.30). The q-correlations are absent if and only if the state of the composite q-system is separable. In other words, if it can be prepared as a statistical mixture of tensor product (uncorrelated) states[6]:

$$\hat{\rho}_{AB} = \sum_\lambda w_\lambda \hat{\rho}_{A\lambda} \otimes \hat{\rho}_{B\lambda}; \quad w_\lambda \geq 0; \quad \sum_\lambda w_\lambda = 1. \tag{4.48}$$

Then and only then there are no q-correlations but classical ones at most. In the contrary case, if $\hat{\rho}_{AB}$ cannot be written in the above form, then the subsystems $A$ and $B$ are said to be in entangled composite state. Accordingly, q-correlation and *entanglement* mean exactly the same thing: the lack of classical separability.

[6] This definition of q-separability was introduced by Werner [5].

The equation of motion of the composite system reads

$$\frac{d}{dt}\hat{\rho}_{AB} = -\frac{i}{\hbar}[\hat{H}_{AB}, \hat{\rho}_{AB}]. \tag{4.49}$$

The composite Hamilton matrix is the sum of the Hamilton matrices of the subsystems themselves plus the interaction Hamilton matrix:

$$\hat{H}_{AB} = \hat{H}_A \otimes \hat{I}_B + \hat{I}_A \otimes \hat{H}_B + \hat{H}_{ABint}. \tag{4.50}$$

If $\hat{H}_{ABint}$ is zero then the tensor product initial state remains tensor product state, the dynamics does not create correlation between the q-subsystems. Non-vanishing $\hat{H}_{ABint}$ does usually create correlation. The motion of the whole q-system is reversible (unitary), of course. But that of the subsystems is not. In case of tensor product initial state $\hat{\rho}_A(0) \otimes \hat{\rho}_B(0)$, for instance, the *reduced q-dynamics* of the subsystem $A$ will represent the time-dependent irreversible q-operation $\mathcal{M}_A(t)$ which we can formally write as

$$\hat{\rho}_A(t) = \mathrm{tr}_B\left[\hat{U}_{AB}(t)\hat{\rho}_A(0) \otimes \hat{\rho}_B(0)\hat{U}^\dagger_{AB}(t)\right] \equiv \mathcal{M}_A(t)\hat{\rho}_A(0). \tag{4.51}$$

The reversibility of the composite state q-dynamics has become lost by the reduction: the final reduced state $\hat{\rho}_A(t)$ does not determine a unique initial state $\hat{\rho}_A(0)$ (cf. Sect. 8.2 for further discussion of reduced q-dynamics).

## 4.6 Collective System

The state (4.3) of a q-system is interpreted on the statistical ensemble of identical systems in the same state. We can form a multiple composite q-system from a big number $n$ of such identical q-systems. This we call collective q-system, its state space is the $n$-fold tensor product of the elementary subsystem's vector spaces:

$$\mathcal{H} \otimes \mathcal{H} \otimes \cdots \mathcal{H} \equiv \mathcal{H}^{\otimes n}. \tag{4.52}$$

The collective state reads

$$\hat{\rho} \otimes \hat{\rho} \otimes \cdots \hat{\rho} \equiv \rho^{\otimes n}, \tag{4.53}$$

while the state vector of a pure collective state is

$$|\psi\rangle \otimes |\psi\rangle \otimes \cdots |\psi\rangle \equiv |\psi\rangle^{\otimes n}. \tag{4.54}$$

If $\hat{A}$ is a q-physical quantity of the elementary subsystem then, in a natural way, one can introduce its arithmetic mean, over the $n$ subsystems, as a collective q-physical quantity

$$\frac{\hat{A} \otimes \hat{I}^{\otimes(n-1)} + \hat{I} \otimes \hat{A} \otimes \hat{I}^{\otimes(n-2)} + \cdots + \hat{I}^{\otimes(n-1)} \otimes \hat{A}}{n}. \tag{4.55}$$

Collective q-physical quantities are not necessarily of such simple form. Their measurement is the collective q-measurement. Contrary to the classical theory, not all collective q-measurements can be reduced to independent measurements on the $n$ subsystems, cf. Sects. 7.1.4 and 10.6.

### *4.6.1 Problems, Exercises*

4.1 *Decoherence-free projective measurement.* There are special conditions to avoid decoherence. Let us prove that the non-selective measurement of a q-physical quantity $\hat{A}$ does not change the measured state $\hat{\rho}$ if and only if $[\hat{A}, \hat{\rho}] = 0$.

4.2 *Mixing the eigenstates.* Let us prove that a state given by the non-degenerate density matrix $\hat{\rho}$ can be prepared by mixing the pure eigenstates of $\hat{\rho}$. What mixing weights shall we use? How must we generalize the method if $\hat{\rho}$ is degenerate?

4.3 *Weak measurement of correlation.* If $\hat{A}$ and $\hat{B}$ are q-physical quantities then $\{\hat{A}, \hat{B}\}$ is another q-physical quantity. We can statistically determine the expectation value $\frac{1}{2}\langle\hat{A}\hat{B} + \hat{B}\hat{A}\rangle$, i.e., the real part of the correlation $\langle\hat{A}\hat{B}\rangle$ if we measure both $\hat{A}$ and $\hat{B}$ after each other, provided at least the first measurement is weak. Let us prove that $\langle\bar{A}B_\lambda\rangle = \frac{1}{2}\langle\hat{A}\hat{B} + \hat{B}\hat{A}\rangle$ where $\bar{A}$ is $\hat{A}$'s weak measurement outcome, $B_\lambda$ is $\hat{B}$'s projective measurement outcome.

4.4 *Separability of pure states.* Let us prove that the pure state $|\psi_{AB}\rangle$ of a composite q-system is separable if and only if it takes the form $|\psi_A\rangle \otimes |\psi_B\rangle$.

4.5 *Unitary cloning?* We could try to duplicate the unknown pure state $|\psi\rangle$, cf. Sect. 5.3, of our q-system by cloning it to replace the prepared state $|\psi_0\rangle$ of another q-system with the same dimension of Hilbert space. Let us prove that the map $|\psi\rangle \otimes |\psi_0\rangle \rightarrow |\psi\rangle \otimes |\psi\rangle$ cannot be unitary. Method: let us test whether the above transformation preserves the value of inner products.

## References


1. von Neumann, J.: Mathematical Foundations of Quantum Mechanics. Princeton University Press, Princeton (1955)
2. Joos, E., Zeh, H.D., Kiefer, C., Giulini, D., Kupsch, K., Stamatescu, I.O.: Decoherence and the Appearance of a Classical World in Quantum Theory, 2nd edn. Springer, Berlin (2003)
3. Aharonov, Y., Albert, D.Z., Vaidman, L.: Phys. Rev. Lett. **60**, 1351 (2008)
4. Diósi, L.: Weak measurements in quantum mechanics. In: Françoise, J.P., Naber, G.L., Tso, S.T. (eds) Encyclopedia of Mathematical Physics, vol. 4. Elsevier, Oxford pp. 276–282 (2006)
5. Werner, R.F.: Phys. Rev. A **40**, 4277 (1989)

# Chapter 5
# Two-State Q-System: Qubit Representations

Obviously the simplest q-systems are the two-state systems. Typical realizations are an atom with its ground state and one of its excited states, a photon with its two polarization states, or an electronic spin with its "up" and "down" states. The smallest unit of q-information, i.e. the qubit, is an abstract two-state q-system. This chapter is technical: you learn standard mathematics of a single abstract qubit.

## 5.1 Computational Representation

The Hilbert space of the two-state q-system is the complex 2-dimensional vector space (4.1). The notion of qubit is best realized in the computational basis. We introduce the computational basis vectors $|0\rangle$ and $|1\rangle$:

$$\{|\mathrm{x}\rangle; \mathrm{x}=0,1\}, \quad \sum_{\mathrm{x}=0,1} |\mathrm{x}\rangle\langle\mathrm{x}| = \hat{I}, \quad \langle\mathrm{x}'|\mathrm{x}\rangle = \delta_{\mathrm{x}'\mathrm{x}}. \tag{5.1}$$

Also the primitive binary q-physical quantity $\hat{\mathrm{x}}$ is defined in the computational basis:

$$\hat{\mathrm{x}} = \sum_{\mathrm{x}=0,1} \mathrm{x}|\mathrm{x}\rangle\langle\mathrm{x}| = |1\rangle\langle 1|. \tag{5.2}$$

This is the (singular) $2 \times 2$ hermitian matrix of the qubit, as q-physical quantity. Its eigenvalues are 0 and 1. Often the q-state, rather than $\hat{\mathrm{x}}$, is called the qubit. The generic pure state is a superposition of the basis vectors:

$$c_0|0\rangle + c_1|1\rangle \equiv \sum_{\mathrm{x}=0,1} c_{\mathrm{x}}|\mathrm{x}\rangle; \quad |c_0|^2 + |c_1|^2 = 1. \tag{5.3}$$

L. Diósi, *A Short Course in Quantum Information Theory*,
Lecture Notes in Physics, 827, DOI: 10.1007/978-3-642-16117-9_5,

This is why we say the qubit carries much richer information than the classical bit does, since the qubit can store the values 0 and 1 in parallel as well.

It is in q-logical operations, see Sects. 6.1.1, 11.8, where the computational basis becomes indispensable. To explore the theoretical structure of a qubit, however, the well known physical representation is more convenient, cf. next section.

## 5.2 Pauli Representation

The mathematical models of all two-state q-systems are isomorphic to each other, regarding the state space and the physical quantities. Also the equations of motion are isomorphic for all closed two-state systems. Therefore the qubit formalism and language can be replaced by the terminology of any other two-state q-system. The electronic spin is the expedient choice. This is a genuine two-state q-system, its formalism is covariant for spatial rotations which guarantees conceptual and calculational advantage.

### 5.2.1 *State Space*

The generic pure state of a two-state q-system takes the following form in a certain orthogonal basis of "up" and "down" states:

$$c_\uparrow|\uparrow\rangle + c_\downarrow|\downarrow\rangle; \quad |c_\uparrow|^2 + |c_\downarrow|^2 = 1. \tag{5.4}$$

An obvious parametrization takes normalization and the free choice of the complex phase into the account:

$$\cos\frac{\theta}{2}|\uparrow\rangle + \mathrm{e}^{\mathrm{i}\varphi}\sin\frac{\theta}{2}|\downarrow\rangle. \tag{5.5}$$

The angular parameters $\theta, \varphi$ can be identified with the standard directional angles of a 3-dimensional real unit vector $\boldsymbol{n}$. This way the above q-state can be parametrized by that unit vector itself:

$$|\boldsymbol{n}\rangle = \cos\frac{\theta}{2}|\uparrow\rangle + \mathrm{e}^{\mathrm{i}\varphi}\sin\frac{\theta}{2}|\downarrow\rangle. \tag{5.6}$$

All pure states of a 2-dimensional q-system have thus been brought to a one-to-one correspondence with the surface points of the 3-dimensional unit-sphere called the Bloch sphere. The two basis q-states $|\uparrow\rangle$, $|\downarrow\rangle$ correspond to the north and south poles, respectively, on the Bloch sphere. Diametric points of the surface will always correspond to a pair of orthogonal q-states: $\langle-\boldsymbol{n}|\boldsymbol{n}\rangle = 0$. Hence $|\pm\boldsymbol{n}\rangle$ form a basis. Moreover, any basis can be brought to this form up to the phases of basis vectors. The modulus of the inner product of any two pure states $|\boldsymbol{n}\rangle, |\boldsymbol{m}\rangle$ is equal to the cosine of the half-angle between the two respective polarization vectors $\boldsymbol{n}, \boldsymbol{m}$:

$$|\langle \boldsymbol{m}|\boldsymbol{n}\rangle| = \cos\frac{\vartheta}{2}, \quad \cos\vartheta = \boldsymbol{mn}. \tag{5.7}$$

For physical reasons, we call the 3-dimensional real Bloch vector $\boldsymbol{n}$ the polarization vector of the pure state $|\boldsymbol{n}\rangle$. As we shall see, the states $|\pm\boldsymbol{n}\rangle$ can be interpreted as the two eigenstates of the corresponding electronic spin-component matrix. As a q-physical quantity, the spin-vector $\hat{\boldsymbol{\sigma}}/2$ of the electron was introduced by Pauli. Without the factor 1/2, we call $\hat{\boldsymbol{\sigma}}$ the vector of polarization. Its Cartesian components are the three Pauli matrices

$$\hat{\sigma}_x = \begin{bmatrix} 0 & 1 \\ 1 & 0 \end{bmatrix}, \quad \hat{\sigma}_y = \begin{bmatrix} 0 & -\mathrm{i} \\ \mathrm{i} & 0 \end{bmatrix}, \quad \hat{\sigma}_z = \begin{bmatrix} 1 & 0 \\ 0 & -1 \end{bmatrix}. \tag{5.8}$$

The generic component of the polarization along the direction $\boldsymbol{n}$ reads

$$\hat{\sigma}_n \equiv n_x\hat{\sigma}_x + n_y\hat{\sigma}_y + n_z\hat{\sigma}_z = \boldsymbol{n}\hat{\boldsymbol{\sigma}}. \tag{5.9}$$

Some basic characteristics of the Pauli matrices are the following:

$$\hat{\sigma}_a = \hat{\sigma}_a^\dagger, \quad \operatorname{tr}\hat{\sigma}_a = 0, \quad [\hat{\sigma}_a, \hat{\sigma}_b] = 2\mathrm{i}\epsilon_{abc}\hat{\sigma}_c, \quad \{\hat{\sigma}_a, \hat{\sigma}_b\} = 2\delta_{ab}\hat{I}. \tag{5.10}$$

Let us identify the former up-down basis states (5.4) by the "spin-up" and "spin-down" eigenstates belonging to the +1 and, respectively, to the $-1$ eigenvalues of the polarization component $\hat{\sigma}_z$:

$$|\uparrow\rangle = \begin{bmatrix} 1 \\ 0 \end{bmatrix}, \quad |\downarrow\rangle = \begin{bmatrix} 0 \\ 1 \end{bmatrix}. \tag{5.11}$$

The polarization $\hat{\sigma}_n$ along the direction $\boldsymbol{n}$ has $|\pm\boldsymbol{n}\rangle$ as eigenstates:

$$\hat{\sigma}_n|\pm\boldsymbol{n}\rangle = \pm|\pm\boldsymbol{n}\rangle. \tag{5.12}$$

This directly implies the spectral expansion of $\hat{\sigma}_n$:

$$\hat{\sigma}_n = |\boldsymbol{n}\rangle\langle\boldsymbol{n}| - |-\boldsymbol{n}\rangle\langle-\boldsymbol{n}|. \tag{5.13}$$

This is why we call and denote $|\boldsymbol{n}\rangle$ as the $n$-up state, while the vector $|-\boldsymbol{n}\rangle$ orthogonal to it we call and denote as the $n$-down state:

$$|\boldsymbol{n}\rangle \equiv |\uparrow n\rangle, \quad |-\boldsymbol{n}\rangle \equiv |\downarrow n\rangle. \tag{5.14}$$

When the reference direction is one of the Cartesian axes, we use the notations like $|\uparrow x\rangle$, $|\downarrow x\rangle$, $|\uparrow y\rangle$, $|\downarrow y\rangle$, while the notations of the distinguished $z$-axis may sometimes be omitted: $|\uparrow z\rangle \equiv |\uparrow\rangle, |\downarrow z\rangle \equiv |\downarrow\rangle$. Typically

$$|\uparrow x\rangle = \frac{1}{\sqrt{2}}\begin{bmatrix}1\\1\end{bmatrix} = \frac{|\uparrow\rangle + |\downarrow\rangle}{\sqrt{2}} \tag{5.15}$$

$$|\uparrow y\rangle = \frac{1}{\sqrt{2}}\begin{bmatrix}1\\\mathrm{i}\end{bmatrix} = \frac{|\uparrow\rangle + \mathrm{i}|\downarrow\rangle}{\sqrt{2}}. \tag{5.16}$$

### 5.2.2 Rotational Invariance

The general form of the 2 × 2 unitary matrices is, apart from an arbitrary complex phase, the following:

$$\hat{U}(\boldsymbol{\alpha}) \equiv \exp\left(-\frac{\mathrm{i}}{2}\boldsymbol{\alpha}\hat{\boldsymbol{\sigma}}\right) = \hat{I}\cos\frac{\alpha}{2} - \mathrm{i}\frac{\boldsymbol{\alpha}\hat{\boldsymbol{\sigma}}}{\alpha}\sin\frac{\alpha}{2}, \tag{5.17}$$

where the real $\boldsymbol{\alpha}$ is called the vector of rotation. To interpret the name, let $R(\boldsymbol{\alpha})$ denote the orthogonal 3 × 3 matrix of spatial rotation around the direction $\boldsymbol{\alpha}$, by the angle $\alpha = |\boldsymbol{\alpha}|$. It can be shown that the influence of the above unitary transformation $\hat{U}(\boldsymbol{\alpha})$ on the state vector is equivalent to the spatial rotation $R(\boldsymbol{\alpha})$ of the polarization vector:

$$\hat{U}(\boldsymbol{\alpha})|\boldsymbol{n}\rangle = |\mathbf{R}^{-1}(\boldsymbol{\alpha})\boldsymbol{n}\rangle, \tag{5.18}$$

$$\hat{U}(\boldsymbol{\alpha})\hat{\boldsymbol{\sigma}}\hat{U}^{\dagger}(\boldsymbol{\alpha}) = \mathbf{R}(\boldsymbol{\alpha})\hat{\boldsymbol{\sigma}}. \tag{5.19}$$

This feature of rotational covariance makes the quick proof of many structural properties possible in the Pauli representation.

### 5.2.3 Density Matrix

The density matrix (4.4) which corresponds to the pure state $|\boldsymbol{n}\rangle$ of a two-state q-system takes this form:

$$|\boldsymbol{n}\rangle\langle\boldsymbol{n}| = \frac{\hat{I} + \boldsymbol{n}\hat{\boldsymbol{\sigma}}}{2}. \tag{5.20}$$

In the distinguished case of $z$-up pure state we write

$$|\uparrow\rangle\langle\uparrow| = \begin{bmatrix}1 & 0\\0 & 0\end{bmatrix} = \frac{\hat{I} + \hat{\sigma}_z}{2}. \tag{5.21}$$

From here, by rotation (5.18, 5.19), we obtain the general equation (5.20). This can be generalized even further. If we allow the polarization vector $\boldsymbol{n}$ to have lengths shorter than 1 we can describe any mixed state as well:

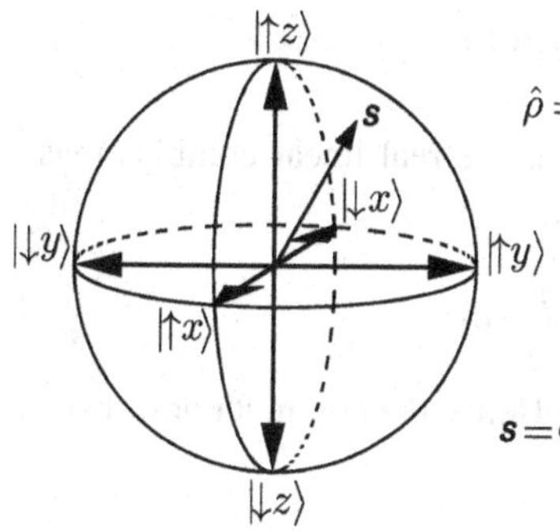


$$\hat{\rho} = \frac{\hat{I} + s\hat{\sigma}}{2} = \frac{1}{2}\begin{bmatrix} 1+s_z & s_x - is_y \\ s_x + is_y & 1 - s_z \end{bmatrix}$$

$|s| \langle 1 \Rightarrow \hat{\rho}$ is mixed state

$|s| = 1 \Rightarrow \hat{\rho}$ is pure state $|s\rangle$

$s = \langle\hat{\sigma}\rangle = \mathrm{tr}\left(\hat{\sigma}\hat{\rho}\right)$ is polarization vector

**Fig. 5.1** Bloch sphere and density matrix. The set of all possible q-states of a qubit can be visualized by the points of a three-dimensional unit sphere of polarization vectors $s$. Surface points ($|s| = 1$) correspond to pure states $|s\rangle$. North and south poles are conventionally identified with $|\uparrow\rangle$, $|\downarrow\rangle$, respectively, whereas the polar coordinates $\theta$, $\varphi$ of the unit vector $s$ coincide with those in the orthogonal expansion (5.5) of pure states. Internal points ($|s| < 1$) correspond to mixed states. The closer $s$ is to the center of the sphere the more mixed is the corresponding state $\hat{\rho}$

$$\hat{\rho} = \frac{\hat{I} + s\hat{\sigma}}{2}; \quad |s| \leq 1. \tag{5.22}$$

The parameter $s$ of the state is just the expectation value of the polarization vector $\hat{\sigma}$, as q-physical quantity, in the q-state $\hat{\rho}$:

$$s = \langle\hat{\sigma}\rangle = \mathrm{tr}(\hat{\sigma}\hat{\rho}). \tag{5.23}$$

### 5.2.4 Equation of Motion

The general Hamilton matrix of a two-state system is the following:

$$\hat{H} = -\frac{1}{2}\hbar\omega\hat{\sigma}, \tag{5.24}$$

where $\omega$ is the vector of external magnetic field provided we identify the system as the electronic spin (and its giro-magnetic factor has been "absorbed" into the scale of the magnetic field). The von Neumann equation of motion (4.5) takes this form:

$$\frac{\mathrm{d}\hat{\rho}}{\mathrm{d}t} = \frac{\mathrm{i}}{2}\omega[\hat{\sigma}, \hat{\rho}]. \tag{5.25}$$

It implies magnetic dipole precession for the expectation value (5.23) of the polarization vector:

$$\frac{\mathrm{d}s}{\mathrm{d}t} = -\omega \times s. \tag{5.26}$$

### 5.2.5 Physical Quantities, Measurement

All q-physical quantities in two-state systems are real linear combinations of the unit-matrix $\hat{I}$ and the three Pauli matrices $\hat{\boldsymbol{\sigma}}$:

$$\hat{A} = \hat{A}^\dagger = a_0\hat{I} + \boldsymbol{a}\hat{\boldsymbol{\sigma}}, \tag{5.27}$$

where $a_0$ is a real number, $\boldsymbol{a}$ is a real vector. Hence the commutator of two such q-physical quantities is

$$[\hat{A}, \hat{B}] = 2\mathrm{i}(\boldsymbol{a} \times \boldsymbol{b}) \cdot \hat{\boldsymbol{\sigma}}, \tag{5.28}$$

in obvious notation. This vanishes only when $\boldsymbol{a}$ and $\boldsymbol{b}$ are parallel vectors. Then, however, the two physical quantities $\hat{A}$ and $\hat{B}$ are simply functions of each other. It is clear that a two-state system has no non-trivial compatible physical quantities Sect. 4.4.4. The maximum compatible set is a single polarization $\hat{\sigma}_n$ in one direction $\boldsymbol{n}$, or the two orthogonal projectors $\hat{P}_{\pm n} = |\pm\boldsymbol{n}\rangle\langle\pm\boldsymbol{n}|$ contributing to its spectral expansion (5.13). The three q-physical quantities $\hat{\sigma}_n, \hat{P}_n, \hat{P}_{-n}$ are functions of each other. The q-measurement of any of them is in all respects equivalent with the measurement of any other one.

Consider a given pure state (5.6)

$$|\uparrow z'\rangle = \cos\frac{\theta}{2}|\uparrow\rangle + \mathrm{e}^{\mathrm{i}\varphi}\sin\frac{\theta}{2}|\downarrow\rangle, \tag{5.29}$$

which is a spin in direction $z'$ of polar angle $\theta$. Suppose the selective measurement of the distinguished polarization $\hat{\sigma}_z$ (4.37, 4.38):

$$|\uparrow z'\rangle \rightarrow \begin{cases} |\uparrow\rangle, & \hat{\boldsymbol{\sigma}}_z \rightarrow +1, \quad p_\uparrow = \cos^2(\theta/2) \\ |\downarrow\rangle, & \hat{\boldsymbol{\sigma}}_z \rightarrow -1, \quad p_\downarrow = \sin^2(\theta/2) \end{cases}. \tag{5.30}$$

We can say that the initial spin of polar angle $\theta$ has jumped into the vertical direction with probability $\cos^2(\theta/2)$. Exploiting the rotational invariance, we resume that a pure state polarized in direction $\boldsymbol{n}$ will, when a certain $\hat{\sigma}_m$ is selectively measured, jump into the new direction $\boldsymbol{m}$ with transition probability

$$p(\boldsymbol{n} \rightarrow \boldsymbol{m}) = |\langle\boldsymbol{m}|\boldsymbol{n}\rangle|^2 = \cos^2(\vartheta/2); \quad \cos\vartheta = \boldsymbol{nm}, \tag{5.31}$$

which is the cosine square of half the angle between the two directions $\boldsymbol{n}$ and $\boldsymbol{m}$. This probability itself turns out to be symmetric between the initial and final states. We shall see in Sect. 6.2.4 of the next chapter that such transition probabilities can be chosen as a measure of likeness—the *fidelity*—between two states in general.

## 5.3 The Unknown Qubit, Alice and Bob

Because of the irreversibility of q-measurements an unknown q-state represents an issue much different from that of an unknown classical state. For various purposes in the forthcoming chapters, we need the precise notion of the unknown qubit. Interestingly, there is no theoretical standard for the totally unknown mixed q-states but for the subset of the pure states. In this sense define we the unknown (random) qubit as a pure state $|\boldsymbol{n}\rangle$ whose polarization $\boldsymbol{n}$ is totally random over the $4\pi$ solid angle. It should not be confused with the totally mixed state $\hat{\rho} = \hat{I}/2$ which would correspond to the average state of the random qubits.

When we notice it, the concept of unknown qubit may be used in a particular sense. We can suppose more knowledge and less ignorance. To make the definitions more operational, it is common to personalize the one who knows the state and the other one who does not. For instance, we shall assume that Alice prepares and gives Bob a qubit in one of the two non-orthogonal states

$$|\uparrow z\rangle \quad \text{or} \quad |\uparrow x\rangle, \tag{5.32}$$

between which she has decided upon her tossing a coin. That is all that also Bob knows but he does not know which of the two states he has actually received. We say that Bob has an unknown qubit in this sense. If interested, he must perform a q-measurement.

## 5.4 Relationship of Computational and Pauli Representations

The computational 5.1 and the Pauli 5.2 representations can be transformed into each other by the conventional mapping of the corresponding bases[1]:

$$|0\rangle = |\uparrow\rangle = \begin{bmatrix} 1 \\ 0 \end{bmatrix}, \quad |1\rangle = |\downarrow\rangle = \begin{bmatrix} 0 \\ 1 \end{bmatrix}. \tag{5.33}$$

The elemental qubit physical quantity (5.2) is in simple algebraic relationship with the polarization component $\hat{\sigma}_z$:

$$\hat{\mathrm{x}} = \frac{\hat{I} - \hat{\sigma}_z}{2}; \quad \hat{\sigma}_z = \hat{I} - 2\hat{\mathrm{x}}. \tag{5.34}$$

The density matrices of the computational basis states can also be expressed by the polarization $\hat{\sigma}_z$:

[1] Perhaps physicists would prefer the other way around: $|\downarrow\rangle = |\downarrow\rangle$ and $|1\rangle = |\uparrow\rangle$.

$$|0\rangle\langle 0| = \frac{\hat{I} + \hat{\sigma}_z}{2}; \quad |1\rangle\langle 1| = \frac{\hat{I} - \hat{\sigma}_z}{2}. \tag{5.35}$$

The same thing in compact form reads

$$|\mathrm{x}\rangle\langle \mathrm{x}| = \frac{\hat{I} + (1 - 2\mathrm{x})\hat{\sigma}_z}{2}; \quad \mathrm{x} = 0, 1. \tag{5.36}$$

## 5.5 Fock Representation

When qubits are represented by two-state systems with a certain energy gap $\varepsilon$, it makes sense to identify the ground and the excited states by $|0\rangle$ and by $|1\rangle$, respectively. We introduce the matrices of emission and absorption:

$$\hat{a} = |0\rangle\langle 1|, \quad \hat{a}^\dagger = |1\rangle\langle 0|. \tag{5.37}$$

They satisfy the anti-commutation relationship

$$\{\hat{a}, \hat{a}^\dagger\} = \hat{I}. \tag{5.38}$$

The matrix $|1\rangle\langle 1|$ is called the occupation number, in Fock representation it reads

$$\hat{n} = \hat{a}^\dagger \hat{a}. \tag{5.39}$$

The occupation number appears in the form of the Hamiltonian as well:

$$\hat{H} = \epsilon \hat{a}^\dagger \hat{a} = \epsilon \hat{n}. \tag{5.40}$$

Powers of $\hat{a}$ (and $\hat{a}^\dagger$) vanish: $\hat{a}^2 = 0$. The matrices $\hat{a}$ and $\hat{a}^\dagger$ can replace the Pauli matrices according to the following relationships:

$$\hat{\sigma}_x = \hat{a} + \hat{a}^\dagger \quad \hat{\sigma}_y = -\mathrm{i}(\hat{a} - \hat{a}^\dagger) \quad \hat{\sigma}_z = 1 - 2\hat{n}. \tag{5.41}$$

It is also common to denote $\hat{a}, \hat{a}^\dagger$ by $\hat{\sigma}_\pm$, respectively:

$$\hat{a} = \frac{\hat{\sigma}_x + \mathrm{i}\hat{\sigma}_y}{2} = \hat{\sigma}_+ \quad \hat{a}^\dagger = \frac{\hat{\sigma}_x - \mathrm{i}\hat{\sigma}_y}{2} = \hat{\sigma}_-. \tag{5.42}$$

As we see, all physical quantities will be of the form

$$\hat{A} = \hat{A}^\dagger = a\hat{I} + b\hat{n} + c^*\hat{a} + c\hat{a}^\dagger, \tag{5.43}$$

where $a$, $b$ are real, $c$ is complex.

The convenience of the Fock representation manifests itself in q-thermodynamics where, as a matter of fact, the absorption and emission of the q-energy $\epsilon$ become the distinguished transitions of the qubit state, see Chap. 12.

## 5.6 Problems, Exercises

5.1 *Pure state fidelity from density matrices.* Let us show that the cosine rule (5.7) for the inner product $\langle \boldsymbol{m}|\boldsymbol{n}\rangle$ of pure states can be derived in a single step starting from the corresponding two density matrices. Method: let us express $|\langle \boldsymbol{m}|\boldsymbol{n}\rangle|^2$.

5.2 *Unitary rotation for* $|\uparrow\rangle \longrightarrow |\downarrow\rangle$. What can be the rotation vector $\boldsymbol{\alpha}$ that rotates the state $|\uparrow\rangle$ into the orthogonal state $|\downarrow\rangle$? Let us make a simple choice for the rotation axis! Calculate the matrix of the corresponding unitary rotation $\hat{U}(\boldsymbol{\alpha})$ and verify the result.

5.3 *Density matrix eigenvalues and -states in terms of polarization.* Let us express the two eigenvalues and the two eigenvectors (i.e. the spectral expansion) of a density matrix as function of the polarization vector $\boldsymbol{s}$.

5.4 *Magnetic rotation for* $|\uparrow\rangle \longrightarrow |\downarrow\rangle$. Determine the (constant) magnetic field $\boldsymbol{\omega}$ that rotates the electronic spin from state $|\downarrow\rangle$ into state $|\uparrow\rangle$. How long time must the field be switched on?

5.5 *Interrelated q-bit physical quantities.* Write down all pairwise relationships for the matrices $\hat{\sigma}_n, \hat{P}_n, \hat{P}_{-n}$.

5.6 *Mixing non-orthogonal polarizations.* Suppose we mix the pure states $|\uparrow z\rangle$ and $|\uparrow x\rangle$ together at rate 1/3:2/3. Write down the polarization vector $\boldsymbol{s}$ of the resulting mixed state.

# Chapter 6
# One-Qubit Manipulations

Numerous principles and methods of q-theory as well as q-information theory can already be demonstrated on a two-state q-system, i.e., on a single qubit. Entanglement can certainly not, we shall present it on composite systems in Chap. 7. The present chapter teaches the elements of q-state manipulation and various simple instances of the q-informatic approach.

## 6.1 One-Qubit Operations

We present the reversible logical operations on single qubits, whose combinations can represent all unitary one-qubit operations. Then we discuss the example of depolarization q-operation generated by polarization measurements. We learn the notorious polarization reflection which is not a q-operation: by classical analogy it were realizable but it turns out not to be so. Let us note that the combinations of one-qubit unitary operations and projective measurements can not represent all possible one-qubit q-operations. These become available if the qubit forms an interacting composite system with an environmental q-system, cf. Chap. 8.

### *6.1.1 Logical Operations*

In q-physics, coherent operations are reversible, unitary transformations. If the coherent operation is applied to an *unknown* q-state then the state remains invariably unknown after the operation. For qubits, the unitary transformations have been mapped onto three-dimensional rotations (5.17). Now we are going to consider those rotations which play a distinguished role for the logical operations on qubits. The Pauli matrices (5.8) themselves, too, are unitary. They are conventionally

L. Diósi, *A Short Course in Quantum Information Theory*,
Lecture Notes in Physics, 827, DOI: 10.1007/978-3-642-16117-9_6,

denoted by $\hat{X}$, $\hat{Y}$ and $\hat{Z}$ whenever qubit-operations are considered, and the corresponding operations are called *X*-, *Y*-, *Z*-operations.

Among the three (non-trivial) classical one-bit logical operations the NOT is the only reversible one. Hence, only the NOT possesses a q-analogue, namely, the *X* operation. The *Z*-operation and the *H*-operation (Hadamard operation) are further one-qubit operations that have no classical counterparts. The reason is that the *Z*-operation inverts the relative phase between the basis states $|0\rangle$ and $|1\rangle$, while the *H*-operation brings both $|0\rangle$ and $|1\rangle$ into their superpositions. Phases and superpositions of $|0\rangle$ and $|1\rangle$ can not be interpreted classically. The elemental one-qubit logical operations and their influence are then the following:

$$\hat{X} = \begin{bmatrix} 0 & 1 \\ 1 & 0 \end{bmatrix}, \quad \left.\begin{matrix} |0\rangle \\ |1\rangle \end{matrix}\right\} \rightarrow \boxed{X} \rightarrow \left\{\begin{matrix} |1\rangle \\ |0\rangle \end{matrix}\right. \tag{6.1}$$

$$\hat{Z} = \begin{bmatrix} 1 & 0 \\ 0 & -1 \end{bmatrix}, \quad \left.\begin{matrix} |0\rangle \\ |1\rangle \end{matrix}\right\} \rightarrow \boxed{Z} \rightarrow \left\{\begin{matrix} |0\rangle \\ -|1\rangle \end{matrix}\right. \tag{6.2}$$

$$\hat{H} = \frac{1}{\sqrt{2}}\begin{bmatrix} 1 & 1 \\ 1 & -1 \end{bmatrix}, \quad \left.\begin{matrix} |0\rangle \\ |1\rangle \end{matrix}\right\} \rightarrow \boxed{H} \rightarrow \left\{\begin{matrix} \frac{1}{\sqrt{2}}(|0\rangle + |1\rangle) \\ \frac{1}{\sqrt{2}}(|0\rangle - |1\rangle) \end{matrix}\right. . \tag{6.3}$$

Like all coherent (unitary) operations, also the above ones are reversible. All three one-qubit operations are identical to their inverses, respectively.

The *X*, *Z*, *H* operations can be combined, e.g. $\hat{X}\hat{Z}\hat{H}\hat{X}$ is a combined unitary operation. Whether all unitary one-qubit operations can be realized by combining a few basic ones? The above three operations are not sufficient. Extend, nonetheless, the half-turn around the *z*-axis (*Z*) by the rotation at an arbitrary angle $\varphi$, called the phase-operation

$$\hat{T}(\varphi) = \begin{bmatrix} e^{-i\varphi/2} & 0 \\ 0 & e^{i\varphi/2} \end{bmatrix}, \quad \left.\begin{matrix} |0\rangle \\ |1\rangle \end{matrix}\right\} \rightarrow \boxed{T(\varphi)} \rightarrow \left\{\begin{matrix} e^{-i\varphi/2}\,|0\rangle \\ e^{+i\varphi/2}\,|1\rangle \end{matrix}\right. \tag{6.4}$$

Now, consider the combined operation $\hat{H}\hat{T}(\theta)\hat{H}$ which is a rotation of the Bloch sphere around the *x*-axis by the angle $\theta$. Let us follow the basis state $|0\rangle$ under the subsequent transformations $\hat{H}, \hat{T}(\theta)$ and $\hat{H}$:

$$|0\rangle \rightarrow \frac{|0\rangle + |1\rangle}{\sqrt{2}} \rightarrow \frac{e^{-i\theta/2}\,|0\rangle + e^{-i\theta/2}\,|1\rangle}{\sqrt{2}} \rightarrow \cos\frac{\theta}{2}\,|0\rangle - i\sin\frac{\theta}{2}\,|1\rangle. \tag{6.5}$$

The resulting state corresponds to a Bloch vector with polar coordinates $\theta$ and $-\pi/2$. We do a further rotation by an angle $\pi/2$ plus $\varphi$ around the *z*-axis. The cumulative influence of the four logical q-operations is the following:

$$\hat{T}\left(\frac{1}{2}\pi + \varphi\right)\hat{H}\hat{T}(\theta)\hat{H}\,|0\rangle = e^{-i(\varphi/2)-i(\pi/4)}\left(\cos\frac{\theta}{2}|0\rangle + e^{-i\varphi}\sin\frac{\theta}{2}|1\rangle\right). \tag{6.6}$$

Apart from an irrelevant complex phase, the r.h.s. shows the generic pure state (5.6) of a qubit. Furthermore, it can be proved that by the combinations of the operations $\hat{H}$ and $\hat{T}(\varphi)$ one can perform all one-qubit unitary operations. Therefore we call the $\hat{H}$ and the $\hat{T}(\varphi)$ a set of *universal* one-qubit operations. This becomes important for the theory of q-computation in Chap. 11.

### 6.1.2 Depolarization, Re-polarization, Reflection

The q-measurement is the most important operation among the irreversible ones. Its irreversibility is related to that we have acquired *hidden* q-information from a q-state and we can not completely re-supply it into the state, because of decoherence. More general irreversible operations can be obtained if we combine q-measurements and unitary operations. It is important that the mapping of the density matrix remains linear and information on the state, cf. Chap. 10 for details, can only be obtained at the expense of irreversibility and decoherence.

We shall see later in Sect. 6.2.5 that there is a naive measurement strategy to determine a single unknown qubit. This naive strategy is just a single measurement of polarization $\hat{\sigma}_n$ along a random direction $\boldsymbol{n}$. The average post-measurement state of the qubit becomes mixed. For a general state, the post-measurement polarization becomes 1/3 smaller than the original polarization:

$$\hat{\rho} \equiv \frac{\hat{I} + \boldsymbol{s}\hat{\boldsymbol{\sigma}}}{2} \to \mathcal{M}\hat{\rho} \equiv \frac{\hat{I} + \boldsymbol{s}\hat{\boldsymbol{\sigma}}/3}{2}. \tag{6.7}$$

The random polarization measurement leads to isotropic depolarization. The corresponding operation is obviously linear.

Consider now the opposite case and suppose that we wish to bring an unknown partially polarized state into a totally polarized one, keeping the direction of polarization invariant:

$$\hat{\rho} \equiv \frac{\hat{I} + \boldsymbol{s}\hat{\boldsymbol{\sigma}}}{2} \to \frac{\hat{I} + \boldsymbol{s}\hat{\boldsymbol{\sigma}}/|\boldsymbol{s}|}{2}, \text{ i.e.: } \boldsymbol{s} \to \frac{\boldsymbol{s}}{|\boldsymbol{s}|}. \tag{6.8}$$

This would mean non-linear transformation for the polarization vector as well as for the density matrix. Hence the above re-polarization of the unknown qubit would contradict the foundations of q-theory and, as a consequence, the corresponding operation does not exist (Sect. 4.3).

Consider finally a map $\mathcal{T}$, also called spin inversion or reflection, which is doubtlessly linear:

$$\hat{\rho} \equiv \frac{\hat{I} + \boldsymbol{s}\hat{\boldsymbol{\sigma}}}{2} \to \mathcal{T}\hat{\rho} \equiv \frac{\hat{I} - \boldsymbol{s}\hat{\boldsymbol{\sigma}}}{2}, \text{ i.e.: } \boldsymbol{s} \to -\boldsymbol{s}. \tag{6.9}$$

This map reflects the polarization of an unknown state. On pure states this map is $|\boldsymbol{n}\rangle \to |-\boldsymbol{n}\rangle$. Such apparently innocent operation, which would classically mean the mere inversion of the electron's angular motion, does not exist in q-physics, it can not be realized. The reason is that, in general, an isolated q-system is correlated with the other q-systems of the world by classical *and* q-correlations. These q-correlations, i.e. these entanglements, will cause that not all linear maps are real operations on the space of q-states. For such inhibited operations the above linear map represents a typical instance. However, the impossibility of the operation can only be explained later in Sect. 7.2.2 by the mechanism of entanglement which assumes composite systems.

## 6.2 State Preparation, Determination

We start with standard applications and proceed to the paradoxical "no-cloning" theorem of q-theory [1]. From this unusual property we derive two applications where q-physics guarantees such security of information processing that classical physics can never do.

### *6.2.1 Preparation of Known State, Mixing*

A certain demanded state, e.g., the pure state $|\uparrow\rangle$ polarized in the $z$-direction, can be prepared from an unknown state if we simply measure the polarization $\hat{\sigma}_z$. If the raw state was a totally random pure state then in half of the cases the measurement outcome is +1 and the resulting state is the demanded $|\uparrow\rangle$. On average, in the other half of the cases the outcome is $-1$. Then we rotate the polarization of $|\downarrow\rangle$ by 180 degrees via the unitary transformation $\hat{\sigma}_x$ (5.17):

$$|\downarrow\rangle \longrightarrow \hat{\sigma}_x|\downarrow\rangle = |\uparrow\rangle, \tag{6.10}$$

so that again we obtain the desired state $|\uparrow\rangle$. The pure states prepared in the above way on demand will serve for preparation of arbitrarily prescribed mixed states.

A generic mixed state (5.22) can be expanded in this form:

$$\hat{\rho} = \frac{\hat{I} + \boldsymbol{s}\hat{\boldsymbol{\sigma}}}{2} = \frac{1+|\boldsymbol{s}|}{2}|\uparrow\boldsymbol{s}\rangle\langle\uparrow\boldsymbol{s}| + \frac{1-|\boldsymbol{s}|}{2}|\downarrow\boldsymbol{s}\rangle\langle\downarrow\boldsymbol{s}|. \tag{6.11}$$

We can thus mix the desired state $\hat{\rho}$ from $\boldsymbol{s}$-up and $\boldsymbol{s}$-down[1] pure states (5.12). But we can do it from other two or more pure states. Even mixed states can be used as raw material for mixing. Let us find the general rule! The basic Eq. (4.8) of mixing two q-states yields the following form for two-state density matrices (5.22):

$$\hat{I} + \boldsymbol{s}\hat{\boldsymbol{\sigma}} = w_1(\hat{I} + \boldsymbol{s}_1\hat{\boldsymbol{\sigma}}) + w_2(\hat{I} + \boldsymbol{s}_2\hat{\boldsymbol{\sigma}}). \tag{6.12}$$

---

[1] The notations $|\uparrow\boldsymbol{s}\rangle, |\downarrow\boldsymbol{s}\rangle$ mean $|\uparrow\boldsymbol{n}\rangle, |\downarrow\boldsymbol{n}\rangle$, resp., where $\boldsymbol{n} = \boldsymbol{s}/|\boldsymbol{s}|$.

Thus the relationship of the polarization vectors must be this:

$$\mathbf{s} = w_1\mathbf{s}_1 + w_2\mathbf{s}_2. \tag{6.13}$$

The rule can be generalized for multiple weighted mixing:

$$\hat{\rho} = \sum_\lambda w_\lambda \hat{\rho}_\lambda \Longleftrightarrow \mathbf{s} = \sum_\lambda w_\lambda \mathbf{s}_\lambda. \tag{6.14}$$

The polarization vector of the mixture is identical to the weighted mean of polarization of the components. The Bloch vector of the mixture can therefore be anywhere within the convex hull of the components Bloch vectors, depending on the distribution of the weights of mixing. Hence a pure state can never be obtained from mixing any other states. True mixed states can be prepared by mixing pure or even mixed states in infinitely many different ways.

The completely depolarized (rotationally invariant) mixed state, i.e., the ensemble behind it, can be mixed from the $z$-up/down pure states:

$$\hat{\rho} \equiv \frac{1}{2}\hat{I} = \begin{cases} 50\% \, |\uparrow z\rangle \\ 50\% \, |\downarrow z\rangle \end{cases}, \tag{6.15}$$

as well as from the $x$-up/down pure states:

$$\hat{\rho} \equiv \frac{1}{2}\hat{I} = \begin{cases} 50\% \, |\uparrow x\rangle \\ 50\% \, |\downarrow x\rangle \end{cases}. \tag{6.16}$$

Recall Sect. 4.2 that after mixing it is totally impossible to distinguish which way the mixed state was prepared. It is crucial, of course, that mixing always means a probabilistic one. To prepare, e.g., the ensemble (6.15), we can use repeated coin tossing to draw the $z$-up/down states for the mixture (Table 6.1).

### 6.2.2 Ensemble Determination of Unknown State

The determination of a completely unknown (not necessarily pure) qubit state $\hat{\rho}$ is equivalent to the determination of the polarization vector $\mathbf{s}$. The expectation value of the physical quantity $\mathbf{s}$ parametrizes the state $\hat{\rho}$:

**Table 6.1** Uniqueness of state, non-uniqueness of mixing

| |
|---|
| The mixed state is uniquely defined by its density matrix $\hat{\rho}$. For a given mixed state (i.e.: for the corresponding ensemble) there no longer exists any test to distinguish which way the mixing was done or, which way the mixed state was altogether created. |
| All is the same for classical densities $\rho$. There is a cardinal departure, however. The decomposition of a classical state (ensemble) into the mixture of pure states is always unique. Differently from the classical mixed states, a non-pure q-state can be decomposed into various mixtures of pure q-states. |

$$\hat{\rho} = \frac{\hat{I} + s\hat{\boldsymbol{\sigma}}}{2}, \qquad s = \mathrm{tr}(\hat{\boldsymbol{\sigma}}\hat{\rho}) = ? \tag{6.17}$$

The three components of $\hat{\boldsymbol{\sigma}}$ are the three Pauli matrices (5.8). One must measure them on the given state $\hat{\rho}$. Since they do not commute, i.e., are not compatible, one must measure them in separate measurements. If the q-ensemble representing the state $\hat{\rho}$ consists of $N$ elements then it is possible to consume cca. $N/3$ elements for the measurement of each Pauli matrix. The outcome of each measurement is either +1 or $-1$; up or down in other words. The obtained statistics allows for the estimation of the polarization components:

$$s_x \approx \frac{N_{\uparrow x} - N_{\downarrow x}}{N_{\uparrow x} + N_{\downarrow x}}, \qquad s_y \approx \frac{N_{\uparrow y} - N_{\downarrow y}}{N_{\uparrow y} + N_{\downarrow y}}, \qquad s_z \approx \frac{N_{\uparrow z} - N_{\downarrow z}}{N_{\uparrow z} + N_{\downarrow z}}, \tag{6.18}$$

and, from them, we can write down the estimation of $\hat{\rho}$. The statistical error is concomitant of the method, it would only disappear at infinite $N$.

The above method is far from being optimum. The problem is open because, first of all, there is no natural definition of "completely" unknown q-state. The situation is simpler if we know at least that the state is pure, i.e. it can be described by a certain state vector $|\boldsymbol{n}\rangle$. The unknown pure state has the natural definition: we suppose it is the random complex unit vector in the Hilbert space. For two-state systems, equivalently, we supposed in Sect. 5.3 that the polarization unit-vector $\boldsymbol{n}$ is a perfectly random unit vector on the Bloch sphere.

### *6.2.3 Single State Determination: No-Cloning*

An unknown q-state can only be determined if we have access to a large number of systems in the same state. If we possess but a single system we would first try copying it, multiplying it. If multiplication were successful, we would produce any number of clones and using this ensemble we would determine the state. In two-state case, we would deliberately reduce the statistical error of polarization estimation.

In reality, however, an unknown q-state can not be cloned. In order of an indirect proof, let us consider a completely mixed q-ensemble of two-state systems. Draw an element of it at random and suppose that we can copy and multiply it in $N$ examples. If the ensemble had been mixed from $z$-up/down states (6.15) then we have $N$ copies either of $z$-up or $z$-down states in our hands. For very large $N$, this method allows us to explore, with high reliability, which states the depolarized ensemble had been mixed from. In particular, the alternative $x$-up/down states can be excluded. This would, however, contradict to the principle that a q-ensemble is uniquely and exhaustively determined by its density matrix.

In summary, the unknown q-state of a single physical system can never be implemented on the other single system so that it, *too*, bear the same q-state. As a consequence, the q-information hidden in the unknown state of a single system can never be copied to the other single system's q-state as well. No carbon copy, no emergency copy can be produced ever.

Suppose, for instance, Alice prepares a single q-system in a pure state. If she does not tell Bob of the direction of polarization then she may safely trust the q-system to Bob since he will never be able to make a copy for himself. If Bob makes an attempt, the copy will usually be faint and Alice will notice the fraud when she gets back either her copy or the fake one, see Sect. 6.2.5.

### 6.2.4 Fidelity of Two States

When a given q-state $\hat{\rho}$ is approximated by another state $\hat{\rho}'$ we may need to measure the quality of the approximation. This purpose is served by the fidelity $F(\hat{\rho}, \hat{\rho}')$. If both states are pure then the definition is this:

$$F = |\langle\psi|\psi'\rangle|^2 = \mathrm{tr}(\hat{\rho}\hat{\rho}'), \tag{6.19}$$

i.e., their fidelity is equal to the squared modulus of their inner product. Fidelity has simple statistical interpretation for pure states. Suppose, for instance, that Bob asks Alice for a certain pure state $|\psi\rangle$, but she passes another pure state $|\psi'\rangle$ instead. When Bob, by the projective measurement of $|\psi\rangle\langle\psi|$, verifies whether he has received the demanded state $|\psi\rangle$ he will find the received state faithful to the demanded one just with the probability $F$.

For qubits $|\boldsymbol{n}\rangle, |\boldsymbol{m}\rangle$, fidelity becomes equal to the squared-cosine of the half-angle between the two polarization vectors:

$$F = |\langle\boldsymbol{m}|\boldsymbol{n}\rangle|^2 = \cos^2(\theta/2); \quad \cos(\theta) = \boldsymbol{mn}. \tag{6.20}$$

### 6.2.5 Approximate State Determination and Cloning

Let us calculate the average fidelity between an unknown qubit $|\boldsymbol{n}\rangle$ and its faint copy $|\boldsymbol{m}\rangle$. Suppose Alice passes a state $|\boldsymbol{n}\rangle = |\uparrow\rangle$ to Bob. Bob needs a copy. He does not know the state and, for lack of a better idea, he measures the polarization $\hat{\sigma}_{z'}$ at a randomly chosen angle $z'$:

$$|\uparrow\rangle \to \begin{cases} |\uparrow z'\rangle, & \hat{\boldsymbol{\sigma}}_{z'} \to +1, \quad p_\uparrow = \cos^2(\theta/2) \\ |\downarrow z'\rangle, & \hat{\boldsymbol{\sigma}}_{z'} \to -1, \quad p_\downarrow = \sin^2(\theta/2) \end{cases} \tag{6.21}$$

and he prepares his copy $|\boldsymbol{m}\rangle$ after the resulting state $|\uparrow z'\rangle$ or $|\downarrow z'\rangle$ according to the measurement outcome $\pm$ 1. The average fidelity of the copy is

$$p_\uparrow|\langle\uparrow|\uparrow z'\rangle|^2 + p_\downarrow|\langle\uparrow|\downarrow z'\rangle|^2. \tag{6.22}$$

Since Bob could only choose the direction $z'$ at random, the expected fidelity of the copy will be the isotropic average over all directions, that is 2/3. What does it mean? If Alice gets the copy back to check it, obviously she measures $\hat{\sigma}_z$ and expects the outcome $\hat{\sigma}_z \to +1$. The probability of this result is just equal to the above calculated fidelity. Hence Alice confirms the copy with 66% probability whereas with 33% probability she obtains the measurement outcome $\hat{\sigma}_z \to -1$ meaning that 1/3 of the copies unveils as completely wrong. Prior to Alice's test, however, it makes no sense to say that the copy is completely true or completely flawed.

## 6.3 Indistinguishability of Two Non-Orthogonal States

In a particular special case of single state determination, one knows a priori that the unknown state is drawn from two known states. The strategy is exceptionally simple when the two states are orthogonal pure states, e.g., we have to decide between

$$|{\uparrow}n\rangle \text{ or } |{\downarrow}n\rangle, \tag{6.23}$$

where $\boldsymbol{n}$ is known. Then a single measurement of $\hat{\sigma}_n$ provides the answer. If, on the contrary, the possible states are not orthogonal, e.g.:

$$|{\uparrow}z\rangle \text{ or } |{\uparrow}x\rangle, \tag{6.24}$$

then there, of course, does not exist any physical quantity such that its single measurement would provide the safe answer. Between two non-orthogonal states it is impossible to decide with full certainty. This is what q-banknote 6.4.1 and q-cryptography 6.4.2 will be based on. Decision of limited reliability is still possible. In general, the single measurement strategy can be optimized for various aspects like, e.g., a higher fidelity 6.2.4, a higher ratio of perfect conclusion 6.3.2, or increased accessible information 10.4.

### 6.3.1 Distinguishing Via Projective Measurement

Let our single qubit have either of the two states $|{\uparrow}z\rangle, |{\uparrow}x\rangle$ at 50–50% apriori likelihoods. We can try a polarization measurement either in direction $z$ or $x$, can make a choice at equal rate. At 75% likelihood the outcome will be 1 and at 25% the outcome will be $-1$. In the former case the measurement is not conclusive. In the latter case the measurement is perfectly conclusive: if we happened to measure $\hat{\sigma}_z = -1$ then the pre-measurement state could not be the $|{\uparrow}z\rangle$ but $|{\uparrow}x\rangle$. And if we measured $\hat{\sigma}_x = -1$ then the pre-measurement state must have been the $|{\uparrow}z\rangle$. The probability of conclusive answer is 25% only. By using non-projective measurement, a higher ratio of conclusive answers can be achieved [2], as it is shown in Sect. 6.3.2.

### 6.3.2 Distinguishing Via Non-projective Measurement

Let us introduce the following positive decomposition (4.12):

$$\hat{\Pi}_z = w|\downarrow x\rangle\langle\downarrow x|, \quad \hat{\Pi}_x = w|\downarrow z\rangle\langle\downarrow z|, \quad \hat{\Pi}_? = \hat{I} - \hat{\Pi}_z - \hat{\Pi}_x. \tag{6.25}$$

If the non-projective measurement yields $\hat{\Pi}_z \to 1$ or $\hat{\Pi}_x \to 1$ then the pre-measurement state must have been $|\uparrow z\rangle$ or $|\uparrow x\rangle$, respectively. If the outcome is $\hat{\Pi}_? \to 1$ then no information is gained. The likelihood of such inconclusive cases follows from (4.20):

$$p_? = \mathrm{tr}\left[\hat{\Pi}_? \frac{|\uparrow z\rangle\langle\uparrow z| + |\uparrow x\rangle\langle\uparrow x|}{2}\right] = 1 - \frac{w}{2}, \tag{6.26}$$

provided the two non-orthogonal states (6.24) had equal apriori probabilities. The non-projective measurement becomes optimal for the maximum weight $w$:

$$w = \frac{\sqrt{2}}{1+\sqrt{2}}, \tag{6.27}$$

i.e., when the effect $\hat{\Pi}_?$ becomes semi-definite:

$$\hat{\Pi}_z = \frac{\sqrt{2}}{1+\sqrt{2}}|\downarrow x\rangle\langle\downarrow x|, \quad \hat{\Pi}_x = \frac{\sqrt{2}}{1+\sqrt{2}}|\downarrow z\rangle\langle\downarrow z|, \quad \hat{\Pi}_? = \frac{2}{1+\sqrt{2}}|\uparrow n\rangle\langle\uparrow n|, \tag{6.28}$$

where

$$\boldsymbol{n} = \frac{(1,0,1)}{\sqrt{2}}. \tag{6.29}$$

Now the ratio of conclusive cases is

$$1 - p_? = \frac{w}{2} = 1 - \frac{1}{\sqrt{2}} \approx 30\%. \tag{6.30}$$

Hence the optimum non-projective measurement has increased the ratio of correct decisions by cca. 5% with respect to projective measurements.

## 6.4 Applications of No-Cloning and Indistinguishability

The non-clonability 6.2.3 of an unknown q-state seems a shortcoming literally. Yet it becomes a benefit if we take a new point of view, namely, the security of information. It can be verified in simple terms that at suitable circumstances the hidden q-information may offer absolute security such that is guaranteed in q-theory whereas classically it can by no means be guaranteed in principle either. The q-theory of unforgeable banknote had been the earliest example. The secure q-key distribution is, for the time being, the only *q-protocol* used already in practice.

### 6.4.1 Q-Banknote

Suppose that Alice, on the behalf of the bank of issue, makes a prepared two-state quantum system stick to each issued banknote, each in randomly different q-state which is of course logged by Alice and the bank. The bank keeps the logs private. Civilians like Bob have no access to the data. Suppose Bob attempts to duplicate or even multiplicate a banknote. Since the maximum fidelity of the copies is 2/3, cf. Sect. 6.2.5, the bank will, on average, unveil 1/3 of the forgeries while 2/3 of the forged money could be perfectly consumed. Just for this reason, as a matter of course, Alice makes more than one independent qubit-marker stick to each banknote. $N$ qubits reduces Bob's chances from 2/3 to $(2/3)^N$.

For Alice and the bank, a simple q-protocol may be the following. Each banknote is identified by the usual public serial number and by a private q-serial-number. The latter is this. Each banknote contains a sequence of $N$ independent two-state q-systems (qubits). Each qubit is either $|0\rangle$ or $|1\rangle$ at random.[2] Hence, each banknote carries a random sequence of alternative pure states, e.g.:

$$|1\rangle, |0\rangle, |0\rangle, \ldots, |0\rangle, |1\rangle, |1\rangle. \tag{6.31}$$

Alice and the bank log the otherwise public serial number together with the encoded $N$-digit binary "q-serial-number". The latter as well as the logs are kept private. At the same time, both states $|0\rangle, |1\rangle$ can be published. If they were orthogonal to each other, Bob would make any number of perfect forges even without destroying the original banknotes. The point is that Alice and the bank should use non-orthogonal states,[3] e.g. (6.24):

$$|0\rangle = |\uparrow z\rangle, \quad |1\rangle = |\uparrow x\rangle, \tag{6.32}$$

then Bob can never profit from forgery. Even doing his best, Bob is not able to produce a sufficient number of high fidelity copies. The bulk of the copied and the original banknotes can not go through a test performed by Alice to verify the logged combination of the public and q-serial numbers. The systematic proof needs detailed statistical analysis which must extend to more sophisticated cloning strategies which Bob might invent. Let us add that prior to Alice's test, it makes no

---

[2] Despite identical notations, one should not confuse $|0\rangle, |1\rangle$ in this chapter with the computational basis in the rest of the volume.

[3] This idea of Wiesner from cca. 1969 remained unpublished for many years until [3].

sense to say that a single banknote is bad or good. This is hidden q-information as long as the test is not yet done.

The task of longtime coherent preservation of arbitrary qubits is not yet solved. Therefore q-banknotes have so far not been realized since the qubit states ought to be preserved on the banknote for an unlimited time. The no-cloning principle has nonetheless another application which is realizable easier, as we see in Sect. 6.4.2.

### 6.4.2 Q-Key, Q-Cryptography

The fact of non-clonability can be utilized in secret, protected from the unauthorized, communication of information. In this case the coherent preservation of a qubit must be guaranteed for not longer than the duration of its transmission. This duration is particularly short when qubits are transmitted by light.

The simplest classical cryptography is based on a secret-key. This may be a sequence of binary digits which is known exclusively by the authorized parties (Alice and Bob) who will utilize the key to encode secret messages to each other. In course of distribution of the key between the authorized parties it may go through the hands of the unauthorized person (Eve) and she might unnoticedly read and copy the key. Even if Eve does not learn the exact key she can still break the secrecy of the communication between Alice and Bob. Q-cryptography excels classical cryptography in that the *unnoticed* eavesdropping is impossible during the distribution of the secret key. The collaborative procedure of Alice and Bob in order to establish the secret key is governed by the so-called secret-key q-protocol. Let us see the simplest two-state q-protocol.[4]

Alice sends a long sequence of random binary digits (raw-key) to Bob, encoded into $N$ non-orthogonal qubits like in case of q-banknotes (6.31, 6.32). Alice also allows Bob (and anybody else) to know the two non-orthogonal states $|0\rangle$ and $|1\rangle$, like e.g. in (6.24), that she has used to encode 0 and 1 respectively. Accordingly, Bob measures each qubit in turn. For lack of a better idea, he alternates randomly between the measurement of the two physical quantities

$$\hat{\sigma}_z \equiv 2|0\rangle\langle 0| - \hat{I} \text{ or } \hat{\sigma}_x \equiv 2|1\rangle\langle 1| - \hat{I}. \tag{6.33}$$

The potential outcomes of the measurements are always $\pm 1$. The average frequency of the $-1$ is 25%, and in all these cases Bob obtains exact information about the qubit received from Alice. When, for instance, the measurement of $\hat{\sigma}_z$ yields $-1$ then the received state only can have been the $|\uparrow x\rangle$, never the $|\uparrow z\rangle$. Hence Bob can safely establish that Alice's raw-key contains the 0 in the given binary digit. Similarly, if the outcome of the measurement of $\hat{\sigma}_x$ was $-1$ then Bob establishes that the given digit of the key is 1. In such a way Bob restores 1/4 of the

4 This is Bennett's version BB92 [4] of the original four-state BB84 protocol [5] by Bennett and Brassard. For a review on q-cryptography, cf. Gisin et al. [6].

raw-key, this is the sift-key. Then Bob allows Alice (and anybody else) to know which digits of the raw-key have contributed to the sift-key. The above protocol supplies Alice and Bob with the same binary sequence of cca. $N/4$ digits, which they can use as the secret-key of their cryptographic communication.

In the language of communication theory, Alice sends the qubits, encoding the raw-key, through a *q-channel* to Bob. Further communications of the protocol, like that describing the code qubits to Bob and the locations of the bits of the sift-key to Alice, are sent through classical channels. Both the classical and q-channels are public, anybody has the access to the bits or q-bits travelling through them. Yet, the resulting sift-key is protected from unnoticed eavesdropping. Eve can unnoticedly eavesdrop the classical channel but she does not obtain any useful knowledge regarding the sift-key. She must concentrate on the q-channel and intercept some of the qubits. If Eve eavesdrops the q-channel then Alice and Bob will detect it because their sift-keys will not be identical. Surely, Eve can not discriminate two non-orthogonal states without altering the original state at the same time. Eve can certainly bar Alice and Bob from establishing the secret key if, e.g., she eavesdrops the q-channel too aggressively so that she may as well make the sift-keys of Alice and Bob completely uncorrelated. In this case there is nothing else left, Alice and Bob re-start the protocol.

In reality, the sift-keys of Alice and Bob may differ because of the eavesdropping by Eve and/or because of the own transmission noise of the q-channel (we ignore the noise during the classical communications of the protocol). Fortunately, Alice and Bob can afterwards eliminate the above differences. They can apply classical algorithms: *information reconciliation* and *privacy amplification*. The former makes, actually, error correction while the latter reduces the relevancy of the information procured by the unauthorized Eve. At the end of the day, Alice and Bob arrive at shorter keys than the original sift ones, which will, nonetheless, coincide with very high reliability while Eve's related information will not exceed a given small value. This happens when the transmission noise, caused by the q-channel itself and/or by Eve's attack on it, is not too high. If it is, then attempts of information reconciliation and privacy amplification will eat up the sift key completely.

A simple method of information reconciliation goes like this. Using classical public communication, Alice and Bob single out randomly the same two locations $k, l$ from their sift-keys. They calculate the parity $\mathrm{x}_k \oplus \mathrm{x}_l$ of the corresponding bits and compare their results using the public classical channel. If the results are different, Alice and Bob eliminates both the $k'$th and the $l'$ bits from their sift-keys. If the results are equal then they agree to drop one of the bits, say the $k'$th so that Eve can not acquire any further information on the key even if she listened to the public classical communication of Alice and Bob. The above procedure is iterated by Alice and Bob until they see that the frequency of disagreement between their shortened keys is already smaller than a given little threshold. If the length of the key is still big enough, then Alice and Bob moves to privacy amplification. At the price of further shortening the key, it reduces the information that may has earlier been acquired by Eve. Once again, in public classical communication, Alice and

**Table 6.2** Two-state q-key protocol

| | | | | | | | |
|---|---|---|---|---|---|---|---|
| Alice's raw key | 1 | 0 | 0 | 1 | 0 | 1 | ... |
| Alice's qubits to Bob | $\|\uparrow x\rangle$ | $\|\uparrow z\rangle$ | $\|\uparrow z\rangle$ | $\|\uparrow x\rangle$ | $\|\uparrow z\rangle$ | $\|\uparrow x\rangle$ | ... |
| Bob's measured quantities | $\hat{\sigma}_z$ | $\hat{\sigma}_x$ | $\hat{\sigma}_x$ | $\hat{\sigma}_z$ | $\hat{\sigma}_z$ | $\hat{\sigma}_x$ | ... |
| Bob's measured outcomes | +1 | −1 | +1 | −1 | +1 | +1 | ... |
| Bob to Alice | | * | | * | | | ... |
| Distributed sift key | | 0 | | 1 | | | ... |

Bob single out randomly the same two locations $k$, $l$ from their keys. They calculate the parity $x_k \oplus x_l$ of the corresponding bits and substitute their original two bits by one new bit which is the parity. By iterating this procedure the information of Eve, obtained earlier, will gradually decrease until Alice and Bob will with high probability have the identical secret keys in their hands, on which Eve (or anyone else) can not have more information than a given (small) limit. We give some hint of the information theoretical proof in Sect. 9.7 (Table 6.2).

## 6.5 Problems, Exercises

1. *Universality of Hadamard and phase operations.* Let us show that all $2 \times 2$ unitary matrices (upto a complex phase) can be constructed by the repeated application of the Hadamard- $\hat{H}$ and the phase-operation $\hat{T}(\varphi)$. Method: use Euler angles of rigid body rotation kinematics.
2. *Statistical error of qubit determination.* Suppose Alice hands over $N$ identically prepared qubits $\hat{\rho}$ to Bob but she does not tell Bob what the state $\hat{\rho}$ is. Let Bob estimate the polarization vector $\boldsymbol{s}$ by the simple method 6.2.2. Write down the estimation errors $\Delta s_x$, $\Delta s_y$, $\Delta s_z$ in function of $\boldsymbol{s}$ and the number of single state measurements. Method: determine the mean statistical fluctuation of the counts $N_{\uparrow x}, N_{\uparrow y}, N_{\uparrow z}$.
3. *Fidelity of qubit determination.* Suppose Alice sends Bob a random qubit $|\boldsymbol{n}\rangle$. Bob knows this but he does not know the state itself. Bob measures a polarization $\hat{\sigma}_{\boldsymbol{m}}$ chosen along a random direction $\boldsymbol{m}$. Let us determine the best expected fidelity of Bob's state estimate.
4. *Post-measurement depolarization.* Alice prepares for Bob a state of polarization $\boldsymbol{s}$ which is unknown to Bob but he wishes to learn it. On the received state, Bob would measure the polarization along a random direction, for lack of a better idea. Let us prove that after the measurement the polarization of the state reduces to $\boldsymbol{s}/3$.
5. *Anti-linearity of polarization reflection.* The polarization reflection $\mathcal{T}$ is equivalent with time inversion and corresponds to anti-linear transformation $\hat{T}$ on the space of state vectors. Let us verify that, upto a joint complex phase factor, the basis vectors transform like $\hat{T}\,|\uparrow\rangle = -|\downarrow\rangle$ and $\hat{T}\,|\downarrow\rangle = |\uparrow\rangle$.

6. *General qubit effects*. Suppose the q-effects

$$\hat{\Pi}_n = w_n\big(\hat{I} + \boldsymbol{a}_n\hat{\boldsymbol{\sigma}}\big); \quad n = 1, 2, \ldots$$

which form a positive decomposition for a qubit, cf. Sect. 4.4. List the necessary conditions on the $w_n$'s and $\boldsymbol{a}_n$'s.

## References

1. Wootters, W.K., Zurek, W.K.: Nature **299**, 802 (1982)
2. Ivanovic, I.D.: Phys. Lett. A **123**, 257 (1987)
3. Wiesner, S.: SIGACT News **15**, 77 (1983)
4. Bennett, C.H.: Phys. Rev. Lett. **68**, 3121 (1992)
5. Bennett, C.H., Brassard, G.: Quantum cryptography: public key distribution and coin tossing, In: Proceedings of IEEE International Conference on Computers, Systems and Signal Processing, IEEE Press, New York (1984)
6. Gisin, N., Ribordy, G., Tittel, W., Zbinden, H.: Rev. Mod. Phys. **74**, 145 (2002)

# Chapter 7
# Composite Q-System, Pure State

Even the simplest q-systems cannot exhaustively be discussed without the concept of composite systems. As we shall see, the reason is q-correlations between otherwise independent q-systems. While classical correlations permit separate treatment of local systems, q-correlations will only permit this with particular limitations. The mathematics of composite q-systems will be introduced from the aspect of q-correlations (entanglements). The reader may learn three historical instances of q-correlation. Two genuine q-informatic applications based on q-correlations will close the chapter.

## 7.1 Bipartite Composite Systems

The simplest composite system has four-states, consists of two 2-state subsystems, i.e., of two qubits. Peculiar features of the composite q-systems can already be interpreted for such $2 \times 2$-state systems as well. Also the major part of our theoretical knowledge on q-correlations has been obtained for $2 \times 2$-state systems. Yet we start with a summary of the general bipartite composite systems, although we restrict the quantitative theory of entanglement for pure composite states

$$|\psi_{AB}\rangle = \sum_{\lambda=1}^{d_A} \sum_{l=1}^{d_B} c_{\lambda l} |\lambda; A\rangle \otimes |l; B\rangle, \tag{7.1}$$

where $\{|\lambda; A\rangle\}$ and $\{|l; B\rangle\}$ are certain orthogonal bases in the respective subsystems $A$ and $B$.

L. Diósi, *A Short Course in Quantum Information Theory*,
Lecture Notes in Physics, 827, DOI: 10.1007/978-3-642-16117-9_7,

### 7.1.1 Schmidt Decomposition

The state vector $|\psi_{AB}\rangle$ of a bipartite composite q-system consists of probability amplitudes $c_{\lambda l}$ of double indices (4.43) where the first (Greek) one labels the basis of subsystem $A$ while the second (Latin) index labels the basis of subsystem $B$. Therefore the state vector can formally be considered a matrix and we can apply the theorem of Schmidt diagonalization to it: the above two orthogonal bases can be chosen in such a way that the matrix $c_{\lambda l}$ of amplitudes becomes diagonal, real, and non-negative. Taking normalization into account, we can write this theorem in the following form:

$$\{c_{\lambda l}\} = \sqrt{w_\lambda}\delta_{\lambda l}, \ \lambda, l = 1, 2, \ldots \min\{d_A, d_B\}. \tag{7.2}$$

Observe that we have formally written the normalized non-negative amplitudes as the square-roots of a normalized probability distribution $w_\lambda$ whose interpretation becomes clear below. The Schmidt decomposition generalizes the diagonal expansion of hermitian matrices, cf. Sect. 4.4, for arbitrary, even non-quadratic matrices. The rank of the given matrix coincides with the number of the nonzero diagonal terms, called also the Schmidt number. Hence, according to the Schmidt decomposition, the pure state of a bipartite composite q-system can always be written as the superposition of orthogonal tensor product state vectors

$$|\psi_{AB}\rangle = \sum_\lambda \sqrt{w_\lambda}\,|\lambda; A\rangle \otimes |\lambda; B\rangle, \tag{7.3}$$

where the number of terms is the rank of the matrix of the composite state amplitudes. This is at most the dimension of the "smaller" subsystem and in this way it may be "much" less than the dimension of the composite system.

The reduced states of the two subsystems follow:

$$\begin{aligned} \hat{\rho}_A &\equiv \mathrm{tr}_B(|\psi_{AB}\rangle\langle\psi_{AB}|) = \sum_\lambda w_\lambda\,|\lambda; A\rangle\langle\lambda; A|, \\ \hat{\rho}_B &\equiv \mathrm{tr}_A(|\psi_{AB}\rangle\langle\psi_{AB}|) = \sum_l w_l\,|l; B\rangle\langle l; B|. \end{aligned} \tag{7.4}$$

Observe that the bases and the coefficients of the Schmidt decomposition can be obtained from the eigenvectors and eigenvalues of the reduced density matrices $\hat{\rho}_A, \hat{\rho}_B$. Their spectra $\{w_\lambda\}$ and $\{w_l\}$ are identical, in this sense the mixednesses of $\hat{\rho}_A$ and $\hat{\rho}_B$ are identical. (Do not forget, this is only true when the state of the composite system is pure.) The reduced states remain pure if and only if the state vector of the composite state is of the tensor product form $|\psi_{AB}\rangle = |\psi_A\rangle \otimes |\psi_B\rangle$, i.e., the Schmidt number is 1.

### 7.1.2 State Purification

An arbitrary mixed state $\hat{\rho}$ of a q-system can be derived by reduction from a pure state of a suitably constructed fictitious larger composite q-system. Let us form the composite system consisting of the q-system in question and a fictitious *environmental* q-system $E$.[1] Consider the spectral expansion (4.11, 4.36) of the state $\hat{\rho}$ of the original q-system:

$$\hat{\rho} = \sum_{\lambda} w_{\lambda} |\lambda\rangle\langle\lambda|. \tag{7.5}$$

Introducing a basis for the environmental q-system $E$, we form the following pure state of the composite ("big") system:

$$|\psi_{\text{big}}\rangle = \sum_{\lambda} \sqrt{w_{\lambda}} |\lambda\rangle \otimes |\lambda; E\rangle, \tag{7.6}$$

where we assume that the dimension of $E$ is not less than the dimension of the original system. By the way, the above form is a Schmidt decomposition (7.3). If we reduce it, we recover the density matrix (7.5) of the original q-system:

$$\hat{\rho} = \text{tr}_E(|\psi_{\text{big}}\rangle\langle\psi_{\text{big}}|). \tag{7.7}$$

Consequently, any mixed state $\hat{\rho}$ can be considered as the reduction of the pure state $|\psi_{\text{big}}\rangle$ of an enlarged q-system. The procedure as well as the state $|\psi_{\text{big}}\rangle$ are called purification of $\hat{\rho}$. An alternative axiomatic construction of q-theory can therefore be started from pure states instead of mixed ones.

One must avoid a certain terminological inexactitude. Purification of q-states is not a q-operation to be performed on the given q-ensemble. Rather it is a mathematical construction, a nonlinear map.

### *7.1.3 Measure of Entanglement*

Separability of composite q-systems has been defined by (4.48): the density matrix must be a mixture of uncorrelated (tensor product) states. For pure composite states this is only possible if the state vector is of the tensor product form

$$|\psi_{AB}\rangle = |\psi_A\rangle \otimes |\psi_B\rangle. \tag{7.8}$$

Tensor product state vectors yield obviously tensor product density matrices which are separable. But non-product state vectors are entangled. This follows from the fact that, in the separability condition (4.48), the decomposition of a pure composite state $\hat{\rho}_{AB}$ must reduce to a single tensor product density matrix

$$\hat{\rho}_{AB} \equiv |\psi_{AB}\rangle\langle\psi_{AB}| = \hat{\rho}_A \otimes \hat{\rho}_B, \tag{7.9}$$

and this can only be satisfied if the state vector $|\psi_{AB}\rangle$ has the product form (7.8).

[1] Sometimes we call them the principal system and the ancilla, or the system and the meter in case of the indirect measurement 8.3.

Accordingly, the pure composite q-state is either of the tensor product form and is thus completely uncorrelated (separable) or, alternatively, it is the superposition of tensor products and is thus correlated (entangled). Hence, for a pure composite q-state, correlations are always q-correlations, and their structural source is the superposition of uncorrelated (tensor product) state vectors. Therefore it makes sense to define the measure $E$ of entanglement on the Schmidt decomposition (7.3). We postulate that the entanglement measure be equal to the Shannon entropy 9.1 of the distribution $w_\lambda$ governing the structure of superposition:

$$E(|\psi_{AB}\rangle\langle\psi_{AB}|) = -\sum_{\lambda} w_\lambda \log w_\lambda, \ \ 0 \leq E \leq \log\min\{d_A, d_B\}. \tag{7.10}$$

If $E = 0$ then all $w_\lambda$ vanish except for a single one which is unity; the composite state vector becomes a tensor product which is, indeed, not entangled. In the contrary case, $E$ takes its maximum value when the distribution $w_\lambda$ is flat:

$$|\psi_{AB}\rangle = \frac{1}{\sqrt{\min\{d_A, d_B\}}}\sum_{\lambda} |\lambda; A\rangle \otimes |\lambda; B\rangle, \tag{7.11}$$

and this state we consider maximally entangled. In this extremal case we shall see that the composite q-system is asymptotically equivalent with $E = \log\min\{d_A, d_B\}$ pairs of maximally entangled qubits in Sect. 7.1.6. The more intrinsic meaning of the postulated relationship (7.10) between entropy and entanglement can be verified later, in the possession of the rudiments of information theory in Sects. 10.5, 10.6, 10.7. Still formally, we introduce the von Neumann entropy 10.1 of an arbitrary q-state:

$$S(\hat{\rho}) = -\mathrm{tr}(\hat{\rho}\log\hat{\rho}); \quad 0 \leq S \leq \log d. \tag{7.12}$$

The von Neumann entropy is zero for pure states and may serve as a measure of the mixedness of the state. But we use it here to express the measure $E$ of entanglement of a given composite pure state $|\psi_{AB}\rangle$. The von Neumann entropies of the reduced states $\hat{\rho}_A, \hat{\rho}_B$ (7.4) of the two subsystems coincide with each other and with the previously defined (7.10) measure of entanglement:

$$E(|\psi_{AB}\rangle\langle\psi_{AB}|) = S(\hat{\rho}_A) = S(\hat{\rho}_B). \tag{7.13}$$

In summary, the entanglement of a pure bipartite composite q-state is measured by the (coinciding) von Neumann entropies of the reduced q-states of the two subsystems. For mixed bipartite composite q-states the criterion of separability as well as the measure of entanglement are much more difficult and only partially known (cf. Table 7.1). There exist separable but classically correlated states; this can only happen to mixed composite states, never to pure ones.

### 7.1.4 Entanglement and Local Operations

How can we create entanglement? How can we bring a product state vector into a superposition of such products? If we choose the mere dynamical way (4.49), then we need a nonzero interaction Hamiltonian (4.50). If there were no interaction then the product initial state vector would remain of the product form. The entanglement $E$ (7.13) of a general initial state vector would not change either! Why, in the lack of their interaction the subsystems $A$ and $B$ would evolve independently of each other, according to their own respective Hamilton matrices $\hat{H}_A, \hat{H}_B$. Their unitary evolution leaves their entropies unchanged.

By means of suitable dynamical interaction, the entanglement of $A$ and $B$ can of course be created, modified, increased, decreased or just eliminated. However, in particular q-informatic situations we may not assume dynamical interaction between the two q-subsystems because, for instance, they are far distant from each other. Commonly, we talk about the q-subsystems $A$ and $B$ as the system of Alice and Bob, respectively, who can only influence their own *local* systems by the given choice of the local Hamilton matrices $\hat{H}_A, \hat{H}_B$. Accordingly, we say that the Hamiltonian is *local* if $\hat{H}_{ABint} = 0$ and is nonlocal otherwise. Usually Alice and Bob are assumed to use local dynamics so they cannot influence the entanglement between their systems in the dynamical way.

Another elemental mean of influencing a given q-state is when we measure a certain q-physical quantity. In case of bipartite composite systems, we call a physical quantity *local* if its measurement can be performed via suitable local measurements by Alice and Bob, respectively. Quantities $\hat{A} \otimes \hat{I}$ or $\hat{I} \otimes \hat{B}$ are typical local ones. The following quantity is typically nonlocal:

$$\hat{A} \otimes \hat{B} + \hat{A}' \otimes \hat{B}'. \tag{7.14}$$

By means of local measurements, Alice and Bob are still able to determine its expectation value but the post-measurement state will be different from what it would be after the standard q-measurement of the nonlocal quantity itself. Its q-measurement is not at all viable by local q-measurements.

It will be shown in Sect. 8.5 of next chapter that local operations (LO) can never create or increase entanglement. Prior to the proof, we need to determine the notion of the most general operations 8.3. If, however, there is already some entanglement then it can by local methods concentrated onto a given composite subsystem. The protocol, quite unusual in common q-theory, is called distillation, Sect. 10.6.

### 7.1.5 Entanglement of Two-Qubit Pure States

If both $A$ and $B$ are 2-state q-systems (qubits) in pure states and are uncorrelated then their composite (two-qubit) system is a pure state of the product form

$$|\psi_{AB}\rangle = |\mathbf{n}_A\rangle \otimes |\mathbf{n}_B\rangle. \tag{7.15}$$

However, the general pure two-qubit state is entangled since it can be written as the superposition of four mutually orthogonal uncorrelated states like this, e.g., in Pauli representation (5.11):

$$c_{\uparrow\uparrow}|\uparrow\rangle \otimes |\uparrow\rangle + c_{\downarrow\uparrow}|\downarrow\rangle \otimes |\uparrow\rangle + c_{\uparrow\downarrow}|\uparrow\rangle \otimes |\downarrow\rangle + c_{\downarrow\downarrow}|\downarrow\rangle \otimes |\downarrow\rangle. \tag{7.16}$$

The same thing in computational (5.33) representation reads

$$\sum_{\mathrm{x}_1=0,1}\sum_{\mathrm{x}_2=0,1} c_{\mathrm{x}_1,\mathrm{x}_2}|1\rangle \otimes |\mathrm{x}_2\rangle \equiv \sum_{\mathrm{x}=0}^{3} c_{\mathrm{x}}|\mathrm{x}\rangle, \quad \mathrm{x} \equiv \mathrm{x}_1\mathrm{x}_2. \tag{7.17}$$

As we see, introducing the two-digit binary number x, i.e.: $2\mathrm{x}_1 + \mathrm{x}_2$, assigns integer labels 0, 1, 2, 3 to the four basis vectors and this allows a compound notation of the general pure state in $2 \times 2$ dimensions.

So far we have considered bases of the two-qubit system constructed as the tensor product of the respective bases of the two single-qubit systems. Non-product bases can equally be useful. In the theory of composite spins, the following four orthogonal basis states are common:

$$\text{singlet: } \frac{|\uparrow\downarrow\rangle - |\downarrow\uparrow\rangle}{\sqrt{2}}, \tag{7.18}$$

$$\text{triplet: } |\uparrow\uparrow\rangle, \frac{|\uparrow\downarrow\rangle + |\downarrow\uparrow\rangle}{\sqrt{2}}, |\downarrow\downarrow\rangle. \tag{7.19}$$

We used compact notations like $|\uparrow\downarrow\rangle$ for $|\uparrow\rangle \otimes |\downarrow\rangle$. The singlet and the middle one of the triplet states are maximally entangled (7.11) while the other two triplet states are uncorrelated. The singlet state is rotationally invariant. This can be made explicit if we write down its density matrix[2]

$$\hat{\rho}(\text{singlet}) = \frac{|\uparrow\downarrow\rangle - |\downarrow\uparrow\rangle}{\sqrt{2}} \frac{\langle\uparrow\downarrow| - \langle\downarrow\uparrow|}{\sqrt{2}} = \frac{\hat{I}_{AB} - \hat{\boldsymbol{\sigma}}_A \otimes \hat{\boldsymbol{\sigma}}_B}{4}. \tag{7.20}$$

Accordingly, the states $|\uparrow\rangle, |\downarrow\rangle$ in the expression of the singlet state vector (7.18) can be equivalently replaced by $|\uparrow n\rangle, |\downarrow n\rangle$ no matter what $n$ is.

In the Bell basis [1], all four orthogonal vectors are maximally entangled:

$$\begin{aligned} |\Phi^\pm\rangle &= \frac{|\uparrow\uparrow\rangle \pm |\downarrow\downarrow\rangle}{\sqrt{2}}, \\ |\Psi^\pm\rangle &= \frac{|\uparrow\downarrow\rangle \pm |\downarrow\uparrow\rangle}{\sqrt{2}}. \end{aligned} \tag{7.21}$$

---

[2] The factors of two $\hat{\sigma}$'s, when appear as $\hat{\boldsymbol{\sigma}} \otimes \hat{\boldsymbol{\sigma}}$ or $\hat{\boldsymbol{\sigma}}\hat{\rho}\hat{\boldsymbol{\sigma}}$, etc., should be understood as spatial scalar products. E.g.: $\hat{\boldsymbol{\sigma}} \otimes \hat{\boldsymbol{\sigma}} = \hat{\sigma}_x \otimes \hat{\sigma}_x + \hat{\sigma}_y \otimes \hat{\sigma}_y + \hat{\sigma}_z \otimes \hat{\sigma}_z$.

The polarizations are completely correlated in the first two basis states and completely anti-correlated in the second two. The Bell basis vector $|\Psi^-\rangle$ coincides with the singlet state (7.20). Another Bell basis vector takes a compound form in computational basis (7.17):

$$|\Phi^+\rangle = \frac{1}{\sqrt{2}} \sum_{\mathrm{x}=0,1} |\mathrm{x}\rangle \otimes |\mathrm{x}\rangle. \tag{7.22}$$

Each of the four Bell states has entanglement $E = 1$ by definition (7.13). We use two-qubit systems in Bell states as standard components to build or to decompose large entangled bipartite systems, as we shall see in 7.1.6, 10.6, 10.7.

### *7.1.6 Interchangeability of Maximal Entanglements*

It is desirable to show that a general bipartite system in pure state of entanglement $E$ is equivalent with $E$ independent copies of two-qubit systems each in Bell states of entanglement 1. We present the proof for the special case of maximum entanglement. For the general pure entangled state, the proof 10.6, 10.7 will need further q-informatic notions.

Suppose a maximally entangled bipartite pure state whose Schmidt decomposition (7.11) reads

$$|\psi_{AB}\rangle = \frac{1}{\sqrt{d}} \sum_{\lambda=1}^{d} |\lambda; A\rangle \otimes |\lambda; B\rangle, \tag{7.23}$$

where, for simplicity, let $d_A = d_B = d$ and let $E = \log(d) \equiv k$ be integer. Alice's $d$-state system can be built up from $k$ qubits, i.e., in the form of a $k$-fold composite system of 2-state subsystems. A straightforward correspondence between the two bases can be obtained by the $k$-digit binary decomposition of the label $\lambda$ of the original basis vectors:

$$|\lambda; A\rangle = |\mathrm{x}_1; A\rangle \otimes |\mathrm{x}_2; A\rangle \otimes \cdots \otimes |\mathrm{x}_k; A\rangle; \quad \lambda = \mathrm{x}_1\mathrm{x}_2, \ldots, \mathrm{x}_k \equiv \mathrm{x}. \tag{7.24}$$

Similarly, we introduce another basis in the state space of Bob. In the new bases, the state (7.23) can be written into this form:

$$\begin{aligned} &\frac{1}{\sqrt{d}} \sum_{\mathrm{x}=1}^{d} [|\mathrm{x}_1; A\rangle \otimes |\mathrm{x}_2; A\rangle \otimes \cdots \otimes |\mathrm{x}_k; A\rangle] \otimes [|\mathrm{x}_1; B\rangle \otimes |\mathrm{x}_2; B\rangle \otimes \cdots \otimes |\mathrm{x}_k; B\rangle] \\ &= \frac{1}{\sqrt{d}} \left( \sum_{\mathrm{x}=0,1} |\mathrm{x}; A\rangle \otimes |\mathrm{x}; B\rangle \right)^{\otimes k} = |\Phi^+\rangle^{\otimes k}. \end{aligned} \tag{7.25}$$

In such a way, we have proved that the pure state (7.23) in $d \times d$-dimension, when it possesses maximum entanglement $E = \log(d)$, is in a suitable basis equivalent with $E$ copies of $2 \times 2$-dimensional (Bell) states of maximum entanglements $E = 1$:

$$|\psi_{AB}\rangle = |\Phi^+\rangle^{\otimes E}. \tag{7.26}$$

## 7.2 Q-Correlations History

The appreciation of unusual correlations in composite q-systems as well as their theoretical characterization took a lengthy time historically. In each of the forthcoming three sections, we recall a decisive theoretical discovery.

### *7.2.1 EPR, Einstein Nonlocality 1935*

Suppose that Alice and Bob possess one qubit each. As a result of their earlier interaction, these qubits are being in a maximally entangled state, e.g., in the singlet state of the Bell basis (7.21):

$$|\psi_{AB}\rangle \equiv |\Psi^-\rangle = \frac{|\uparrow\downarrow\rangle - |\downarrow\uparrow\rangle}{\sqrt{2}}, \tag{7.27}$$

where, as usual, the first polarization refers to Alice's and the second refers to Bob's qubit, respectively. Alice and Bob, together with their qubit, live far from each other. Imagine that Alice measures $\hat{\sigma}_z$ on her qubit. This is the random result of the measurement:

$$|\psi_{AB}\rangle \rightarrow \begin{cases} |\uparrow z; \downarrow z\rangle\,; & \hat{\sigma}_{Az} \rightarrow 1; \quad p_+ = 1/2 \\ |\downarrow z; \uparrow z\rangle\,; & \hat{\sigma}_{Az} \rightarrow -1; \quad p_- = 1/2. \end{cases} \tag{7.28}$$

It is clear that the state of Bob's qubit becomes $|\uparrow\rangle$ or $|\downarrow\rangle$ with 50–50% probabilities, perfectly anti-correlated with Alice's post-measurement state. Alternatively, if Alice measures $\hat{\sigma}_x$ instead of $\hat{\sigma}_z$ then Bob's post-measurement state becomes the mixture of $|\uparrow x\rangle$ and $|\downarrow x\rangle$ instead of $|\uparrow\rangle$ and $|\downarrow\rangle$. At a superficial glance, the post-measurement state of Bob's qubit is completely different if Alice measures $\hat{\sigma}_x$ instead of $\hat{\sigma}_z$. In such a way Alice could transmit information to Bob via action-at-a-distance without any physical interaction, merely exploiting the entanglement of their qubits. We know that this cannot be so: Bob has access but his own qubit and its reduced density matrix $\hat{\rho}_B$ is totally independent from whatever measurement made by Alice. In our case $\hat{\rho}_B$ is the maximally mixed state. Its decomposition is never unique. It is really ambiguous, exactly like in (6.15) and (6.16), depending on Alice remote choice $\hat{\sigma}_z$ or $\hat{\sigma}_x$ to measure. Yet Bob can never detect the difference.

There is, however, a crucial lesson. Einstein gave a particular strong formulation to the notion of locality. According to the principle of Einstein locality: In a theory *completely describing* the physical reality, the action on system $A$ cannot influence the *description* of system $B$ if $A$ and $B$ are spatially separated. As we see from the above EPR paradox [2], the q-theory violates Einstein locality. EPR chose the following resolution: the q-theory does not give complete description of the physical reality. Let us discuss the underlying motivations of EPR.

If Alice and Bob shared a correlated *classical* composite system, the local selective measurement (2.16) by Alice would, similarly to the above q-case, influence the state (e.g.: phase–space distribution) of Bob's remote system. This is, however, merely the consequence of the incompleteness of the description: certain parameters of the composite system have remained hidden and treated statistically. In classical physics it is simple to see that the complete description is possible when the system is in pure state, i.e. all canonical variables are exactly specified. Then the description of the classical system becomes deterministic and cannot violate Einstein locality anymore. In the q-theory, however, even pure states can violate Einstein locality. This is why EPR consider q-theory incomplete. Whether q-theory, too, can be made complete if we discover its *hidden parameters* and we specify their values? Whether this complete theory will satisfy Einstein locality? The negative answer comes soon in Sect. 7.2.3.

### 7.2.2 A Non-Existing Linear Operation 1955

Consider the maximally entangled singlet state $|\Psi^-\rangle$ of two qubits, whose density matrix reads

$$\frac{\hat{I}\otimes\hat{I}-\hat{\boldsymbol{\sigma}}\otimes\hat{\boldsymbol{\sigma}}}{4}. \tag{7.29}$$

Like in case of EPR, the first qubit belongs to Alice, the second to Bob, who are far from each other. Define the polarization reflection (6.9) of a generic qubit, which could be a hypothetical linear q-operation $\mathcal{T}$:

$$\mathcal{T}\frac{\hat{I}+\mathbf{s}\hat{\boldsymbol{\sigma}}}{2}=\frac{\hat{I}-\mathbf{s}\hat{\boldsymbol{\sigma}}}{2}. \tag{7.30}$$

Assume that Alice is able to perform such an operation and she does it as well on her qubit entangled in the singlet state. It is straightforward to see the result. The reflection changes the sign of polarization of Alice's own Pauli matrices while it preserves Bob's ones:

$$(\mathcal{T}\otimes\mathcal{I})\frac{\hat{I}\otimes\hat{I}-\hat{\boldsymbol{\sigma}}\otimes\hat{\boldsymbol{\sigma}}}{4}=\frac{\hat{I}\otimes\hat{I}+\hat{\boldsymbol{\sigma}}\otimes\hat{\boldsymbol{\sigma}}}{4}. \tag{7.31}$$

Surprisingly, this $4\times 4$ Hermitian matrix has become indefinite! For instance, its quadratic form with the singlet state is negative:

**Table 7.1** The Peres–Horodecki test of separability

Let $\mathcal{T}$ be the *reflection* of a single qubit:
$\mathcal{T}\frac{\hat{I}+\mathbf{s}\hat{\boldsymbol{\sigma}}}{2}=\frac{\hat{I}-\mathbf{s}\hat{\boldsymbol{\sigma}}}{2}$.
Let $\hat{\rho}$ be a two-qubit state. If its *partial reflection* is non-negative:
$(\mathcal{T}\otimes\mathcal{I})\hat{\rho}\geq 0$,
then the state $\hat{\rho}$ is *separable*. If its partial reflection is indefinite:
$(\mathcal{T}\otimes\mathcal{I})\hat{\rho}\ngeq 0$,
then the state $\hat{\rho}$ is *entangled*.

$$\langle\Psi^-|\frac{\hat{I}\otimes\hat{I}+\hat{\boldsymbol{\sigma}}\otimes\hat{\boldsymbol{\sigma}}}{4}|\Psi^-\rangle=\operatorname{tr}\left(\frac{\hat{I}\otimes\hat{I}-\hat{\boldsymbol{\sigma}}\otimes\hat{\boldsymbol{\sigma}}}{4}\frac{\hat{I}\otimes\hat{I}+\hat{\boldsymbol{\sigma}}\otimes\hat{\boldsymbol{\sigma}}}{4}\right)=-\frac{1}{2}. \tag{7.32}$$

Hence the local reflection (7.30) of the singlet state is no more a density matrix.[3] It can be shown that this anomaly arises not just for the singlet state but for all entangled two-qubit states. Therefore the local (also called partial) reflection makes an elegant mathematical test of entanglement, this is the Peres–Horodecki test [4, 5] (Table 7.1).

The polarization reflection $\mathcal{T}$ cannot be performed in reality. To interpret the reflection (7.30) of a given local qubit, say of Alice, one should assume that it is not entangled with whatever other qubits in the Universe. Such an assumption, however, cannot be justified. On the contrary, there are classical and q-correlations between local and various remote q-systems of the Universe. But while the existence of external classical correlations does not influence the structural properties (the theory) of local systems, the existence of external q-correlations (entanglements) imposes serious conditions for (the theory of) local q-systems.

### 7.2.3 Bell Nonlocality 1964

Can we make the q-theory complete if we introduce certain *hidden parameters* so that Einstein nonlocality 7.2.1 disappear from it? Bell pointed out that, like the classical theory, also the q-theory can be made complete and deterministic via introducing suitable hidden parameters. (From this aspect it is irrelevant that such a theory would be more complicated and less elegant than the standard q-theory.) Bell has, however, also proved that the obtained complete description cannot satisfy the locality principle of Einstein. In Bell's formulation: no local hidden parameter model can replace the q-theory [6].

---

3 This happens just because $\mathcal{T}$ belongs to non-completely positive maps, cf. Chap. 8, discovered by Stinespring [3].

Like in case of EPR nonlocality, we suppose Alice and Bob share an ensemble of singlet states $|\Psi_{AB}\rangle = |\Psi^-\rangle$ (7.27). Let us consider two couples of local physical quantities on the side of Bob and Alice, respectively:

$$\hat{A} = \boldsymbol{a}\hat{\boldsymbol{\sigma}}_A, \quad \hat{A}' = \boldsymbol{a}'\hat{\boldsymbol{\sigma}}_A, \quad \hat{B} = \boldsymbol{b}\hat{\boldsymbol{\sigma}}_B, \quad \hat{B}' = \boldsymbol{b}'\hat{\boldsymbol{\sigma}}_B, \tag{7.33}$$

where $\boldsymbol{a}$, $\boldsymbol{a}'$, $\boldsymbol{b}$, $\boldsymbol{b}'$ are polarization unit vectors. Alice and Bob measure four different nonlocal combinations

$$\hat{A} \otimes \hat{B}, \quad \hat{A}' \otimes \hat{B}, \quad \hat{A} \otimes \hat{B}', \quad \hat{A}' \otimes \hat{B}'. \tag{7.34}$$

The measurement of the expectation values of such nonlocal quantities is possible with local measurements. For instance, the measurement of $\langle \hat{A} \otimes \hat{B} \rangle$ is equivalent with the simultaneous local measurements of $\hat{A}$ and $\hat{B}$ by Alice and Bob, respectively. But the four combinations are not compatible therefore they cannot be simultaneously measured. Alice and Bob will take four independent measurement statistics on four randomly chosen sub-ensembles of the q-ensemble of the singlet states. Alice and Bob measure in coincidence. The outcome of each local measurement is $\pm 1$. Having the four statistics been taken, Alice and Bob can calculate the expectation value of the following nonlocal quantity, cf. also (7.14):

$$\hat{\mathbf{C}} = \hat{A} \otimes \hat{B} + \hat{A}' \otimes \hat{B} + \hat{A} \otimes \hat{B}' - \hat{A}' \otimes \hat{B}'. \tag{7.35}$$

Let us now suppose that there exist hidden parameters which completely determine the outcomes of all measurements. We also adopt a further, delicate and much discussed, assumption: let the hidden parameters determine the outcomes of the chosen measurements as well as what the outcomes *would have been*, if we had chosen to do one of the other q-measurements incompatible with the one that was actually done. Let us then index the configurations of the hidden parameters by $r$. Each q-system of the singlet-state q-ensemble has its $r$. The hidden parameters assign definite outcome $\pm 1$ to each local polarization measurement (7.33):

$$\hat{A} = A_r = \pm 1, \quad \hat{A}' = A'_r = \pm 1, \quad \hat{B} = B_r = \pm 1, \quad \hat{B}' = B'_r = \pm 1. \tag{7.36}$$

Note that such a hidden parameter description of the measured reality would trivially satisfy the Einstein-locality: any manipulation by Alice on her qubit of hidden parameter $r$ would not influence the description, i.e. the values $B_r$, $B'_r$, of Bob's qubit. Hence we call the $r$'s local hidden parameters.

The relative frequencies of the $\pm 1$ outcomes as well as their correlations should, of course, satisfy the predictions of q-theory. For instance:

$$\langle \hat{A} \otimes \hat{B} \rangle = \lim_{N \to \infty} \frac{1}{N} \sum_{r \in \Omega} A_r B_r; \quad N = |\Omega|, \tag{7.37}$$

and similarly for the other three combinations (7.34) as well. This implies the following identity:

$$\langle \hat{\mathbf{C}} \rangle = \lim_{N\to\infty} \frac{1}{N} \sum_r A_r B_r + A'_r B_r + A_r B'_r - A'_r B'_r. \tag{7.38}$$

Consider the important relationship

$$A_r B_r + A'_r B_r + A_r B'_r - A'_r B'_r = \pm 2, \tag{7.39}$$

which is a consequence of (7.36). Therefore the existence of the local hidden parameter expression (7.38) imposes the following restriction on the correlation quantity $\hat{\mathbf{C}}$ (7.35):

$$-2 \le \langle \hat{\mathbf{C}} \rangle \le 2. \tag{7.40}$$

This is the Bell inequality.[4] Note that this inequality itself is purely classical, its derivation has nothing to do with q-theory. We can ask whether it is satisfied by the value $\langle \hat{\mathbf{C}} \rangle$ predicted by the q-theory.

We substitute the combination (7.35) of the polarizations (7.33) in place of $\langle \hat{\mathbf{C}} \rangle$. In singlet state we obtain

$$-2 \le -\boldsymbol{ab} - \boldsymbol{a}'\boldsymbol{b} - \boldsymbol{ab}' + \boldsymbol{a}'\boldsymbol{b}' \le 2. \tag{7.41}$$

This geometrical condition must be satisfied for all possible choices of the four directions. But it will not be. Suppose, for instance, the following four co-planar directions:

$$\boldsymbol{a} = \rightarrow, \; \boldsymbol{a}' = \uparrow, \; \boldsymbol{b} = \swarrow, \; \boldsymbol{b}' = \nwarrow. \tag{7.42}$$

Insert the corresponding four scalar products into the Bell inequality (7.41):

$$\begin{aligned} &-\boldsymbol{ab} - \boldsymbol{a}'\boldsymbol{b} - \boldsymbol{ab}' + \boldsymbol{a}'\boldsymbol{b}' \\ &= -\cos(3\pi/4) - \cos(3\pi/4) - \cos(3\pi/4) + \cos(\pi/4) = 2\sqrt{2}. \end{aligned} \tag{7.43}$$

This value violates the Bell inequality. It means that in singlet state the values (7.36) of the four physical quantities cannot be chosen as a function of the hidden parameters $r$ in such a way that the statistical predictions of the q-theory be satisfied. This is called Bell nonlocality. Separable states are always Bell local. But we do not know whether or not all entangled states show Bell nonlocality [8].

[4] It is the version by Clauser et al. [7].

## 7.3 Applications of Q-Correlations

We learn two direct applications of q-correlations: superdense coding, and teleportation. Options of information manipulations become available which could not be possible in classical physics.

### 7.3.1 *Superdense Coding*

In the basic setup of communication, Alice sends information through a channel to Bob via classical binary data. What can they profit if Alice sends qubits, instead of classical bits, through a q-channel? When Alice and Bob agree that Alice encodes the binary information into orthogonal qubits, e.g. into $|\uparrow\rangle, |\downarrow\rangle$, then Bob measures $\hat{\sigma}_z$ and their q-communication becomes equivalent with the classical communication. But q-communication can be more powerful, at least in solving certain particular tasks.

Suppose then a q-channel from Alice to Bob which, for the time being, they can use without limitation as well as they can communicate through a classical channel without limitation. They know that one year later Alice obtains classical information from somewhere and she must forward it to Bob. They also know that one year later the classical channel is already not available and they can only send qubits through the q-channel. Intuition says that they have to consume one qubit to encode one classical bit. We shall see, however, that Alice and Bob can, well in time, agree upon the protocol of superdense coding which encodes two bits into one qubit [9].

Alice prepares pairs of qubits entangled in singlet states and sends one qubit of each pair to Bob through the q-channel. Both Alice and Bob preserve their own qubits. Of course, Alice neither could nor wished to send any information to Bob via the entangled qubits. One year later, Alice learns the first two bits that she will forward to Bob via one qubit. She encodes the two bits into the four different unitary 1-qubit operations $\hat{I}, \hat{X}, \hat{Y}, \hat{Z}$. She performs the actual operation on the first of her qubits kept with her. Then the whole composite singlet state becomes one of the four orthogonal Bell states (7.21), according to the scheme

$$
\begin{aligned}
(\hat{I} \otimes \hat{I})\,|\Psi^-\rangle &= |\Psi^-\rangle \\
(\hat{X} \otimes \hat{I})\,|\Psi^-\rangle &= |\Phi^-\rangle \\
(\hat{Y} \otimes \hat{I})\,|\Psi^-\rangle &= |\Phi^+\rangle \\
(\hat{Z} \otimes \hat{I})\,|\Psi^-\rangle &= |\Psi^+\rangle.
\end{aligned}
\tag{7.44}
$$

After the unitary operation, Alice sends her qubit through the q-channel to Bob who has thus one of the four orthogonal Bell states in his hands, encoding the two bits of information according to the protocol. The information can be decoded if

Bob performs a projective measurement of the Bell basis. In such a way, Bob acquires the original two qubits. If Alice and Bob repeatedly perform the above protocol, Alice will be able to send $2n$ bit information via $n$ qubits while they have to consume $n$ shared singlet states.

Superdense coding does not mean that we squeeze 2 bits of information into 1 qubit. In fact it means that we can encode 2 bits of information into 2 qubits in such a way that physically we do not touch but one of the qubits. Therefore the other qubit can be transmitted to the other party in advance, say, when the classical information is not yet even known. In this sense can the coding into the retained single qubit be considered denser than what would be available classically. For superdense coding, entanglement has obviously been instrumental.

### *7.3.2 Teleportation*

Transmission of an unknown qubit requires a q-channel. Yet qubit transmission is possible through classical channel if the parties share a reservoir of previously entangled qubits. Suppose a q-channel from Alice to Bob which, for the time being, they can use without limitation as well as they can communicate through a classical channel without limitation. They know that one year later Alice receives unknown qubits from somewhere and she must forward them to Bob. They also know that one year later the q-channel is already not available and they can only send bits through the classical channel. Intuition says that classical communication is not likely to transmit q-information. We shall see, however, that Alice and Bob can, well in time, agree upon the protocol of teleportation [10] which transmits the unknown qubit from Alice to Bob. More precisely, the original qubit remains with Alice all the time while its state becomes perfectly inherited by a raw qubit of Bob. The protocol destroys the state of the original qubit, otherwise teleportation could make perfect cloning, too, which would be a contradiction, cf. Sect. 6.2.3.

Alice prepares pairs of qubits entangled in singlet states and sends one qubit of each pair to Bob through the q-channel. Both Alice and Bob preserves their own qubits. Of course, Alice neither could nor wished to send any information to Bob via the entangled qubits. One year later, Alice receives the first qubit

$$|\psi\rangle = a\,|\uparrow\rangle + b\,|\downarrow\rangle \tag{7.45}$$

that she will forward to Bob via two classical bits. Consider the 3-qubit composite state consisting of the received qubit and a singlet state shared by Alice and Bob:

$$|\psi\rangle \otimes |\Psi^-\rangle = [a\,|\uparrow\rangle + b\,|\downarrow\rangle] \otimes \frac{|\uparrow\downarrow\rangle - |\downarrow\uparrow\rangle}{\sqrt{2}}. \tag{7.46}$$

This can be re-written if we introduce the Bell basis (7.21) for the two qubits on Alice's side:

$$-\frac{1}{2}|\Psi^-\rangle \otimes [a|\uparrow\rangle + b|\downarrow\rangle] + \frac{1}{2}|\Phi^-\rangle \otimes [a|\downarrow\rangle + b|\uparrow\rangle]$$
$$+\frac{1}{2}|\Phi^+\rangle \otimes [a|\downarrow\rangle - b|\uparrow\rangle] - \frac{1}{2}|\Psi^+\rangle \otimes [a|\uparrow\rangle - b|\downarrow\rangle]. \tag{7.47}$$

We see that the singlet state $|\Psi^-\rangle$ on Alice's side is multiplied by the state $a|\uparrow\rangle + b|\downarrow\rangle = |\psi\rangle$ on Bob's side, which is just the state (7.45) received originally by Alice to teleport. In fact all four Bell states on Alice's side are correlated with $|\psi\rangle$ upto the trivial unitary operations $\hat{I}, \hat{X}, \hat{Y}, \hat{Z}$ in turn. Then Alice performs the projective measurement of the Bell basis. There are four possible outcomes $|\Psi^\pm\rangle, |\Phi^\pm\rangle$. Alice transmits 2 bits through the classical channel to inform Bob of the outcome. So Bob learns which transform he must apply to his qubit:

$$\begin{array}{ll} |\Psi^-\rangle: & \hat{I}[a|\uparrow\rangle + b|\downarrow\rangle] \\ |\Phi^-\rangle: & \hat{X}[a|\downarrow\rangle + b|\uparrow\rangle] \\ |\Phi^+\rangle: & \hat{Y}[a|\downarrow\rangle - b|\uparrow\rangle] \\ |\Psi^+\rangle: & \hat{Z}[a|\uparrow\rangle - b|\downarrow\rangle]. \end{array} \tag{7.48}$$

The resulting state of Bob's qubit is precisely $|\psi\rangle$ (7.45) which was the state to teleport. If Alice and Bob repeatedly perform the above protocol, Alice will be able to send $n$ qubits via $2n$ bits while they have to consume $n$ shared singlet states.

The four possible measurement outcomes, obtained by Alice, are always totally random as it follows from the coefficients of the orthogonal decomposition (7.47). Alice does not acquire any information on the qubit that she teleports. Yet she entangles this qubit with the qubit of the shared singlet state, and this entanglement makes then possible that, via 2 bits of classical information, the unknown qubit reappears on Bob's side while it becomes completely smashed on Alice's side. The hidden information on the teleported qubit has been travelled through the chain of entanglements of the 3 qubits involved in the protocol.

The above teleportation protocol applies invariably to mixed states, too. More than that, it teleports all external q-correlations of the given qubit. Suppose that the state to teleport has been entangled with the state of a certain environmental q-system:

$$a|1;E\rangle \otimes |\uparrow\rangle + b|2;E\rangle \otimes |\downarrow\rangle. \tag{7.49}$$

Once the teleportation protocol from Alice to Bob has been done, the environmental state becomes entangled with Bob's state exactly the same way. The proof is algebraically identical with the previous proof of the teleportation of the unentangled state (7.45), if we make the formal substitutions

$$a \to a|1;E\rangle \otimes, \quad b \to b|2;E\rangle \otimes\,. \tag{7.50}$$

Hence Bob's qubit becomes in all respects perfectly identical with the qubit that Alice received to teleport, including all its external q-correlations. The qubit received by Alice is left behind in the completely mixed state.

## 7.4 Problems, Exercises

7.1 *Schmidt orthogonalization theorem.* Let us prove the Schmidt decomposition theorem (7.2) for a complex rectangular matrix $\hat{c}$, starting from the well-known spectral expansion theorem of Hermitian matrices. Method: apply the latter to both matrices $\hat{c}^\dagger\hat{c}$ and $\hat{c}\hat{c}^\dagger$.

7.2 *Swap operation.* Consider two identical q-systems in arbitrary uncorrelated pure state $|\psi\rangle \otimes |\phi\rangle$. The swap matrix is defined by $\hat{S}(|\psi\rangle \otimes |\phi\rangle) = |\phi\rangle \otimes |\psi\rangle$. $\hat{S}$ is unitary and Hermitian. Let us prove that for two qubits

$$\hat{S} = \frac{\hat{I} \otimes \hat{I} + \hat{\boldsymbol{\sigma}} \otimes \hat{\boldsymbol{\sigma}}}{2}.$$

Method: introduce the basis $|\uparrow\uparrow\rangle, |\downarrow\downarrow\rangle, |\uparrow\downarrow\rangle, |\downarrow\uparrow\rangle$.

7.3 *Singlet density matrix.* Let us write down the density matrix of the two-qubit singlet state in Pauli representation. Method: exploit rotational invariance or express $\hat{\rho}(\text{singlet})$ through the swap $\hat{S}$.

7.4 *Local measurement of expectation values.* Let us show that the expectation value of the nonlocal physical quantity $\hat{A} \otimes \hat{B} + \hat{A}' \otimes \hat{B}'$ can be measured locally as well. Method: verify that the expectation values of $\hat{A} \otimes \hat{B}$ and of $\hat{A}' \otimes \hat{B}'$ are locally measurable.

7.5 *Local measurement of certain nonlocal quantities.* A tensor product q-physical quantity $\hat{A} \otimes \hat{B}$ may not be measured locally although its expectation value is always measurable locally. Let us explain why the nonlocal measurement of $\hat{\sigma}_z \otimes \hat{\sigma}_z$ is not equivalent with the local simultaneous measurement of $\hat{\sigma}_z \otimes \hat{I}$ and $\hat{I} \otimes \hat{\sigma}_z$. Let us prove that for the local measurability of $\hat{A} \otimes \hat{B}$ it is sufficient if the spectrum of $\hat{A} \otimes \hat{B}$ is non-degenerate.

7.6 *Nonlocal hidden parameters.* Let us show that the q-theoretic correlations can after all be reproduced by nonlocal hidden parameters. Method: we complete the hidden parameter $r$ by a switch $v = 1, 2, 3, 4$ marking which of the four measuring setups is active. Then we can already assign the values $A_{rv}, B_{rv}, A'_{rv}, B'_{rv}$ suitably to each pair of qubits in the ensemble.

7.7 *Does teleportation clone the qubit?* Teleportation would be a perfect cloner had the procedure not destroyed the state of the original qubit. Let us show that the original qubit is always left behind in the totally mixed state $\hat{\rho} = \hat{I}/2$.

## References

1. Braunstein, S.L., Mann, A., Revzen, M.: Phys. Rev. Lett. **68**, 3259 (1992)
2. Einstein, A., Podolsky, B., Rosen, N.: Phys. Rev. **47**, 777 (1935)
3. Stinespring, W.F.: Proc. Am. Math. Soc. **6**, 211 (1955)

4. Peres, A.: Phys. Rev. Lett. **77**, 1413 (1996)
5. Horodecki, M., Horodecki, P., Horodecki, R.: Phys. Lett. A **223**, 1 (1996)
6. Bell, J.S.: Physics **1**, 195 (1964)
7. Clauser, J.F., Horne, M.A., Shimony, A., Holt, R.A.: Phys. Rev. Lett. **23**, 880 (1969)
8. Popescu, S.: Phys. Rev. Lett. **74**, 2619 (1995)
9. Bennett, C.H., Wiesner, S.J.: Phys. Rev. Lett. **69**, 2881 (1992)
10. Bennett, C.H., Brassard, G., Crépeau, C., Jozsa, R., Peres, A., Wootters, W.K.: Phys. Rev. Lett. **70**, 1895 (1993)

# Chapter 8
# All Q-Operations

So far we have learned certain particular q-operations like the unitary transformations, the q-measurements, and the reduction of state. The set of generic q-operations is much larger. We learn the elegant mathematical classification. We discuss its various physical interpretations typically involving the temporary enlargement of the system by a certain environmental system, subsequent unitary and measurement operations on the obtained composite system, and an ultimate reduction to the original system.

## 8.1 Completely Positive Maps

We call a linear map $\mathcal{M}$ positive if it brings the density matrix $\hat{\rho}$ of a given system into a non-negative matrix whose trace does not exceed 1. In q-theory, we have to define a subset of positive maps which is the completely positive maps.[1] A positive map $\mathcal{M}$ is completely positive if its trivial extension $\mathcal{M} \otimes \mathcal{I}$ for an arbitrary composite system remains positive map. The completely positive maps are always of the following Kraus form (we omit the proof) ( Table 8.1) :

$$\mathcal{M}\hat{\rho} = \sum_n \hat{M}_n \hat{\rho} \hat{M}_n^\dagger, \quad \sum_n \hat{M}_n^\dagger \hat{M}_n \leq \hat{I}, \tag{8.1}$$

and the Kraus forms are always completely positive. However, given a completely positive map $\mathcal{M}$, its Kraus form is never unique: the Kraus matrices $\hat{M}_n$—forming the "sandwich"—can be chosen in many different ways.

There exist non-completely positive maps. Anti-unitary maps are such, like e.g. the reflection $\mathcal{T}$ (7.30) on the two-state q-systems. If we try to construct the

---

[1] Completely positive maps were discovered in Stinespring [1]. For their physical elucidation cf., e.g. Kraus [2].

L. Diósi, *A Short Course in Quantum Information Theory*,
Lecture Notes in Physics, 827, DOI: 10.1007/978-3-642-16117-9_8,

**Table 8.1** Positive and completely positive maps

| The map $\mathcal{M}: \hat{\rho} \to \mathcal{M}\hat{\rho}$ is: | |
|---|---|
| Positive map if | Also completely positive map if |
| $\mathcal{H}$ is Hilbert space | $\mathcal{H}_E$ is arbitrary "environmental" Hilbert space |
| $\hat{\rho}$ is density matrix on $\mathcal{H}$ | $\hat{\rho}_{big}$ is density matrix on $\mathcal{H} \otimes \mathcal{H}_E$ |
| $\mathcal{M}\hat{\rho} \geq 0$ and tr $(\mathcal{M}\hat{\rho}) \leq 1$ | $(\mathcal{M} \otimes \mathcal{I})\hat{\rho}_{big} \geq 0$ |

Kraus representation we fail. Actually, we would get negative sign for some of the "sandwich" terms. The reflection $\mathcal{T}$ can be written typically into this form:

$$\mathcal{T}\hat{\rho} = \frac{1}{2}\hat{\boldsymbol{\sigma}}\hat{\rho}\hat{\boldsymbol{\sigma}} - \frac{1}{2}\hat{\rho}. \tag{8.2}$$

By no means can it be written into the Kraus form (8.1). It is obvious that $\mathcal{T}$ is non-completely positive if we recall Sect. 7.2.2 where we saw that the trivial extension $\mathcal{T} \otimes \mathcal{I}$ for $2 \times 2$ dimension was not a positive map.

The non-completely positive maps can not be realized. As we shall see in the next two sections, the completely positive maps can, at least in principle, be realized in many different ways. What we called q-operations (Sect. 4.2) is nothing else but the completely positive maps.

## 8.2 Reduced Dynamics

Consider the composition of the given q-system and a certain, maybe fictitious, environmental q-system $E$, and introduce a unitary dynamics $\hat{U}$ on the composite system. Let us then reduce the state to the original system, The resulting q-operation is usually non-unitary:

$$\mathcal{M}\hat{\rho} = \mathrm{tr}_E\left[\hat{U}(\hat{\rho} \otimes \hat{\rho}_E)\hat{U}^\dagger\right]. \tag{8.3}$$

The same q-operation $\mathcal{M}$ can be obtained from different unitary dynamics $\hat{U}$, moreover, the choice of the environmental system can also be varied. All trace-preserving q-operations can be obtained as reduced dynamics (8.3). The proof can be done in Kraus representation (8.1). Let us introduce a basis for the environmental system $E$:

$$|n; E\rangle, \quad n = 1, 2, \ldots, d_E. \tag{8.4}$$

Suppose the initial state of $E$ is pure:

$$\hat{\rho}_E = |1; E\rangle\langle 1; E|. \tag{8.5}$$

Then the q-operation (8.3) takes the Kraus form

$$\mathcal{M}\hat{\rho} = \sum_n \hat{M}_n \hat{\rho} \hat{M}_n^\dagger, \tag{8.6}$$

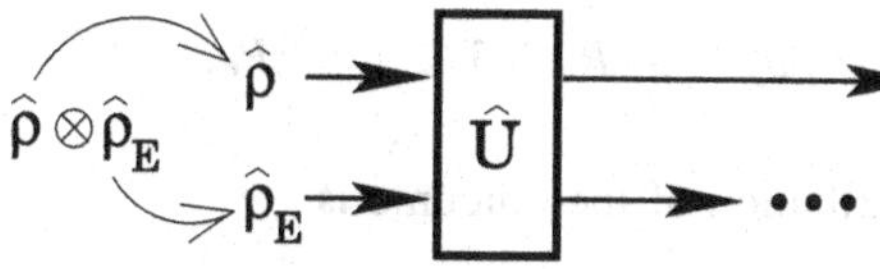


**Fig. 8.1** Reduced q-dynamics vs. q-operations. We let our system in question interact unitarily with a (real or fictitious) environmental system. We suppose the uncorrelated initial state $\hat{\rho}\otimes\hat{\rho}_E$ which the interaction transforms into the q-correlated one $\hat{U}(\hat{\rho}\otimes\hat{\rho}_E)\hat{U}^\dagger$. Finally we ignore the environmental system and we only consider the reduced final state $\mathrm{tr}_E[\hat{U}(\hat{\rho}\otimes\hat{\rho}_E)\hat{U}^\dagger]$. The unitary interaction with the environment has thus induced a non-unitary dynamics for our system, which we call reduced dynamics. It corresponds to a completely positive map $\mathcal{M}$. The inverse relationship is also true: A q-operation $\mathcal{M}$ (i.e.: a completely positive map) can always be realized by a suitable reduced dynamics

if we insert the following Kraus matrices[2]:

$$\hat{M}_n = \mathrm{tr}_E\left[(\hat{I}\otimes|1;E\rangle\langle n;E|)\hat{U}\right] \equiv \langle n;E|\hat{U}|1;E\rangle. \tag{8.7}$$

If, the other way around, we know the Kraus form of a trace-preserving q-operation then the Eq. (8.7) is always solvable for the unitary dynamics $\hat{U}$ if we assume an environmental system $E$ of suitable large dimensions.

## 8.3 Indirect Measurement

Each "sandwich" term of the Kraus form can be interpreted separately. Suppose that we perform a non-selective q-measurement on the environmental system $E$. If we do it *after* the unitary interaction $\hat{U}$ then it does not influence the reduced state of the original system. If, however, our measurement is selective then we can arrive at the individual "sandwich" terms of the Kraus form. To see this, we start from the basis (8.4) of $E$, we form the corresponding partition $\hat{P}_{En} = |n;E\rangle\langle n;E|$ and extend it trivially for the whole composite system:

$$\hat{P}_n = \hat{I}\otimes\hat{P}_{En}. \tag{8.8}$$

Perform the projective measurement of these physical quantities, after the unitary interaction $\hat{U}$. The concept is called indirect measurement. We must note of course that the measurement of the projectors (8.8) is equivalent with the measurement of the projectors $\hat{P}_{En}$ on $E$. Follow the rules of selective measurement (4.13, 4.15). The state of the composite system changes as follows:

[2] To avoid misunderstandings, we note the natural shorthand notation $\langle n;E|\hat{U}|1;E\rangle$ stands for a matrix acting on the states $\hat{\rho}$ of the original system.

$$\hat{\rho} \otimes \hat{\rho}_E \rightarrow \frac{1}{p_n} \hat{P}_n \hat{U} (\hat{\rho} \otimes \hat{\rho}_E) \hat{U}^\dagger \hat{P}_n, \quad (8.9)$$

while the probability distribution of the outcome is

$$p_n = \mathrm{tr}\left[\hat{P}_n \hat{U} (\hat{\rho} \otimes \hat{\rho}_E) \hat{U}^\dagger\right]. \quad (8.10)$$

Now we reduce the first equation to the state of the original q-system; the operation $\mathrm{tr}_E$ yields

$$\hat{\rho} \rightarrow \frac{1}{p_n} \hat{M}_n \hat{\rho} \hat{M}_n^\dagger, \quad (8.11)$$

cf. the definition (8.7) of $\hat{M}_n$. The second equation takes the equivalent form

$$p_n = \mathrm{tr}\left[\hat{M}_n^\dagger \hat{M}_n \hat{\rho}\right]. \quad (8.12)$$

As we see, each "sandwich" term in the Kraus form (8.1) of our q-operation corresponds, apart from a normalizing factor, to what comes out from the selective measurement of the environmental system's basis (8.4). The inverse of the renormalizing factor is always equal to the probability of the given outcome.

We got the key to the interpretation of trace-reducing q-operations. Let us make a selection $\Omega$ of the measurement outcomes and discard the complementary part of the q-ensemble. Such q-operation can be written into this form:

$$\mathcal{M}_\Omega \hat{\rho} = \sum_{n \in \Omega} \hat{M}_n \hat{\rho} \hat{M}_n^\dagger; \quad \sum_{n \in \Omega} \hat{M}_n^\dagger \hat{M}_n < \hat{I}, \quad (8.13)$$

where this time the summation over the index $n$ is only partial. The obtained state is not normalized:

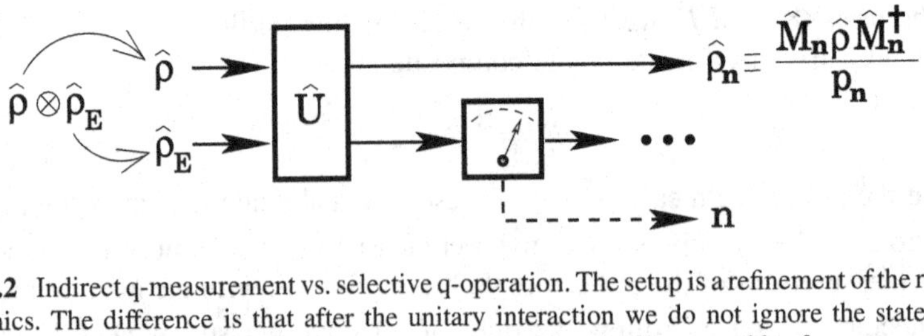


**Fig. 8.2** Indirect q-measurement vs. selective q-operation. The setup is a refinement of the reduced dynamics. The difference is that after the unitary interaction we do not ignore the state of the environment. Rather we perform a projective measurement on it. The ensemble of pre-measurement state $\hat{\rho}$ is selected into sub-ensembles of conditional post-measurement states $\hat{\rho}_n$ according to the obtained measurement outcomes $n$. The probability $p_n$ coincides with the norm of the unnormalized conditional state $\hat{M}_n \hat{\rho} \hat{M}_n^\dagger$. The sub-ensembles of certain conditional post-measurement states $\{\hat{\rho}_n; n \in \Omega\}$ can be re-united contributing to a trace-reducing completely positive map. The inverse relationship is also true: A selective q-operation $\mathcal{M}$ (i.e.: a trace-reducing completely positive map) can always be realized by a suitable indirect measurement

$$\mathrm{tr}[\mathcal{M}_\Omega\hat{\rho}] = \sum_{n\in\Omega} p_n. \tag{8.14}$$

The trace of the state is equal to the total probability of the selected measurement outcomes $n \in \Omega$, i.e., it can be less than 1.

The physical interpretation of q-operations has thus been completed. Each trace-preserving q-operation can be realized as the reduced dynamics (Sect. 8.2) after the suitable unitary interaction with a suitable environmental system. Each trace-compressing q-operation can be realized by a suitable selective indirect measurement via the above environmental system. These realizations are never unique, we can have infinite many choices.

## 8.4 Non-Projective Measurement Resulting from Indirect Measurement

Recall the concept of projective (Sect. 4.4.1) and non-projective (Sect. 4.4.2) q-measurements. Projective measurements are also called von Neumann, standard, or ideal measurements. Non-projective ones, also called non-ideal or unsharp, were formal extensions of projective ones. Now we can easily prove that they are special cases of indirect measurements. Indeed, (8.11, 8.12) will coincide with (4.20, 4.22) of non-projective measurement in the special case of hermitian Kraus matrices. Then we can always identify the effects by the square of the Kraus matrices:

$$\hat{\Pi}_n = \hat{M}_n^2, \qquad \hat{M}_n = \hat{M}_n^\dagger = \hat{\Pi}_n^{1/2}. \tag{8.15}$$

This way, the mechanism of indirect measurements will underly the concept of non-projective measurements. The non-projective measurement of effects $\hat{\Pi}_n$ can always be realized by indirect measurements which are just projective measurements that happen on a certain, maybe fictitious, environmental system $E$ q-correlated with the system in question.

Note that there exists an equivalent projective measurement on the uncorrelated composition $\hat{\rho}\otimes\hat{\rho}_E$ of the system and the environment. Only we have to relax the special form $\hat{P}_n = \hat{I}\otimes\hat{P}_{nE}$ of the projectors and take the unitarily equivalent set

$$\hat{P}_n = \hat{U}^\dagger\hat{I}\otimes\hat{P}_{nE}\hat{U}. \tag{8.16}$$

This time the projective measurement happens on the composite state $\hat{\rho}\otimes\hat{\rho}_E$ as a whole, not only on the environment:

$$\hat{\rho}\otimes\hat{\rho}_E \to \frac{1}{p_n}\hat{P}_n\hat{\rho}\otimes\hat{\rho}_E\hat{P}_n, \tag{8.17}$$

$$p_n = \mathrm{tr}\left[\hat{P}_n(\hat{\rho}\otimes\hat{\rho}_E)\right]. \tag{8.18}$$

Nonetheless, the measurement is equivalent, due to unitary equivalence, with the indirect measurement (8.8–8.10) defined previously. The projective measurement of $\hat{P}_n$ yields the non-projective measurement of the effect $\hat{\Pi}_n$ on the reduced state $\hat{\rho}$. If $\hat{P}_n$ yields 1 or 0 then it means that $\hat{\Pi}_n$ yields 1 or 0 in coincidence. While the former expresses sharp information the latter does not. Projective measurement is repeatable, non-projective is not. When we average over the environmental state we get averaged data. The effects, too, are averaged version of the projectors:

$$\hat{\Pi}_n = tr_E\left[\hat{P}_n(\hat{I}\otimes\hat{\rho}_E)\right]. \tag{8.19}$$

Of course, this expression is equivalent with (8.15).

## 8.5 Entanglement and LOCC

As we saw earlier in Sect. 7.1.4 of the preceding chapter, local q-dynamics and q-measurements can not create q-correlation. Having learned the class of general q-operations (Sect. 8.1), we can now formulate the precise statement. In the postulated situation, Alice and Bob own their separate local systems far from each other and the remote systems form a composite q-system together. Alice and Bob are allowed to perform local operations (LO). When they do so independently then the composite system changes this way:

$$\hat{\rho}_{AB} \longrightarrow (\mathcal{M}_A\otimes\mathcal{M}_B)\hat{\rho}_{AB}. \tag{8.20}$$

Usually we suppose classical communication (CC) as well. Alice and Bob can inform each other about the setups and the outcomes of their local operations so that they can condition their further LOs on the received classical information. Q-correlation (entanglement), if it is not already present, can never be produced by LOCC. To see this, suppose the initial state is uncorrelated $(\hat{\rho}_A\otimes\hat{\rho}_B)$ and Alice performs her LO $\mathcal{M}_A$ selectively, see Eqs. (8.11) and (8.12):

$$\hat{\rho}_A \rightarrow \frac{1}{p_{An}}\hat{M}_{An}\hat{\rho}_A\hat{M}_{An}^\dagger \equiv \hat{\rho}_{An}. \tag{8.21}$$

Suppose that, using CC, Alice tells Bob of the outcome $n$: Bob can make his LO $\mathcal{M}_B$ dependent on $n$. Let $\mathcal{M}_B^{(n)}$ stand for Bob's LO. The selective change of the composite state is this:

$$\hat{\rho}_A\otimes\hat{\rho}_B \rightarrow \hat{\rho}_{An}\otimes\mathcal{M}_B^{(n)}\hat{\rho}_B, \tag{8.22}$$

and the non-selective post-LOCC state becomes:

$$\hat{\rho}_A\otimes\hat{\rho}_B \rightarrow \sum_n p_{An}\hat{\rho}_{An}\otimes\mathcal{M}_B^{(n)}\hat{\rho}_B. \tag{8.23}$$

This is a separable state containing classical correlations only. The proof extends trivially to the case when the initial state is already a mixture of uncorrelated states $(\hat{\rho}_{A\lambda} \otimes \hat{\rho}_{B\lambda})$. Any LOCC will preserve such a separable structure.

## 8.6 Open Q-System: Master Equation

When a q-system is continuously interacting with certain environmental q-system or q-systems then it is called open q-system. A subsystem of a closed system makes a typical open system. The open system dynamics is reduced dynamics and is thus irreversible. There exists an idealized (so-called Markovian) class of environmental interactions where the state would still satisfy an equation similar to von Neumann's (4.5). The Markovian master equation takes this form:

$$\frac{\mathrm{d}\hat{\rho}}{\mathrm{d}t} = -\frac{\mathrm{i}}{\hbar}[\hat{H}, \hat{\rho}] + \text{non-Hamiltonian terms} \equiv \mathcal{L}\hat{\rho}. \tag{8.24}$$

It should generate a completely positive map $\mathcal{M}(t) = \exp(t\mathcal{L})$. This condition implies that the master equation can always be written into Lindblad form [3, 4] (although we do not prove it here):

$$\frac{\mathrm{d}\hat{\rho}}{\mathrm{d}t} \equiv \mathcal{L}\hat{\rho} = -\frac{\mathrm{i}}{\hbar}[\hat{H}, \hat{\rho}] + \sum_n \left( \hat{L}_n \hat{\rho} \hat{L}_n^\dagger - \frac{1}{2}\{\hat{L}_n^\dagger \hat{L}_n, \hat{\rho}\} \right). \tag{8.25}$$

The supermatrix $\mathcal{L}$ is called the Lindblad generator. Given a master equation by $\mathcal{L}$, its Lindblad form is never unique: the Lindblad matrices $\hat{L}_n$—forming the non-Hamiltonian term—can be chosen in many different ways. The physical interpretation of Lindblad matrices is related to the form of the interaction between the open system and its environment. Choice of Lindblad matrices of a given open system is non-unique because the same reduced dynamics (Sect. 8.2) of the open system can be obtained by various interactions with various environments.

## 8.7 Q-Channels

A typical theoretical application of q-operations $\mathcal{M}$ is the concept of a q-channel. A q-channel serves to communicate an arbitrary q-state $\hat{\rho}$ from one location to another, say, from Alice to Bob like in the protocols of the q-banknote (Sect. 6.4.1), q-key distribution (Sect. 6.4.2), or superdense coding (Sect. 7.3.1). In the ideal case, these protocols use noiseless q-channels: the output state coincides with the input one. In general, however, the noisy q-channel will distort the state, it performs a certain q-operation $\mathcal{M}$ on the input. Terminologically, we identify the operation $\mathcal{M}$ and the q-channel. Q-channels are categorized according to the distortions they cause. Elementary q-channels are those communicating single independent qubits. The depolarizing channel is this:

$$\hat{\rho} \rightarrow \mathcal{M}\hat{\rho} = (1-w)\hat{\rho} + w\frac{\hat{I}}{2}, \tag{8.26}$$

it damps the polarization of a qubit isotropically by a factor $1 - w$. The dephasing channel takes this form:

$$\hat{\rho} \rightarrow \mathcal{M}\hat{\rho} = (1-w)\hat{\rho} + w\hat{\sigma}_z\hat{\rho}\hat{\sigma}_z, \tag{8.27}$$

which decreases the off-diagonal elements of $\hat{\rho}$ in the computational basis by a factor $1 - w$.

## 8.8 Problems, Exercises

8.1 *All q-operations are reductions of unitary dynamics.* Consider a trace-preserving q-operation $\mathcal{M}$. It is equivalent with a certain reduced dynamics where the system interacts unitarily with a "fictitious" environmental system of suitably large dimension $d_E$. Let us prove this equivalence starting from the Kraus representation of $\mathcal{M}$ and show that it suffices if we choose $d_E$ to coincide with the number of the Kraus matrices $\hat{M}_n$. Method: introduce the bases $|\lambda\rangle, |n;E\rangle$ for the system and the environment, respectively, and observe that the states $\sum_n \hat{M}_n|\lambda\rangle \otimes |n;E\rangle$ form an orthonormal set.

8.2 *Non-projective effect as averaged projection.* Let us write the uncorrelated composite state of the system and the environmental system as $\hat{\rho}\hat{\rho}_E$. We can many times omit the notation of $\otimes$ as well as $\hat{I}$ without risk of misunderstanding. In this example we use such convenience. Suppose we perform a projective measurement of the partition $\{\hat{P}_n\}$ of the composite system. If we ignore the post-measurement state of the environment, we are left with a given non-projective measurement of the positive decomposition $\{\hat{\Pi}_n\}$ on the system alone. Let us prove that $\hat{\Pi}_n = \mathrm{tr}_E(\hat{P}_n\hat{\rho}_E)$.

8.3 *Q-operation as supermatrix.* Consider an operation $\mathcal{M}$ which transforms density matrices into density matrices. In a given basis, the density matrices are represented by their components $\rho_{\lambda\mu} = \langle\lambda|\hat{\rho}|\mu\rangle$ and the operation is represented by the components of a supermatrix of four indices:

$$\rho_{\lambda\mu} \longrightarrow \sum_{\lambda'\mu'} \mathcal{M}_{\lambda\mu\lambda'\mu'}\rho_{\lambda'\mu'}.$$

Let us express the supermatrix $\mathcal{M}$ by utilizing the Kraus representation.

8.4 *Environmental decoherence, time-continuous depolarization.* If we are interested in the complete isolation of our q-system then all environmental interactions are considered as parasitic. They tend to destroy superpositions and entanglements within our system. They will paralyze, in particular, q-protocols and q-algorithms that take long time enough to accumulate the above environmental decoherence. A simplest model of an uncontrollable environmental interaction is the time-continuous depolarization of the qubit. Let us show that the master equation

$$\frac{\mathrm{d}\hat{\rho}}{\mathrm{d}t} = -\frac{1}{4\tau}(\hat{\boldsymbol{\sigma}}\hat{\rho}\hat{\boldsymbol{\sigma}} - 3\hat{\rho})$$

is of the Lindblad form and it describes isotropic depolarization. Derive the equation of motion for the polarization vector $\boldsymbol{s}$.

8.5 *Kraus representation of depolarization*. Let us construct Kraus representation for the depolarization channel. Method: exploit the isotropy of the channel.

## References


1. Stinespring, W.F.: Proc. Am. Math. Soc. **6**, 211 (1955)
2. Kraus, K.: States, Effects, and Operations: Fundamental Notions of Quantum Theory. Springer, Berlin (1983)
3. Lindblad, G.: Commun. Math. Phys. **48**, 199 (1976)
4. Gorini, V., Kossakowski, A., Sudarshan, E.C.G.: J. Math. Phys. **17**, 821 (1976)

# Chapter 9
# Classical Information Theory

Shannon entropy is the key-notion of classical information. It provides the statistical measure of information associated with states $\rho$. Since dynamical aspects shall not be treated at all, we would just talk about probability distributions $p$ instead of physical states $\rho$. For comparability with Q-information theory of Chap. 10, however, we keep talking about states $\rho$ of classical systems. Typically, we use heuristic proofs though corner stones of the exact derivations will fairly be indicated.

## 9.1 Shannon Entropy, Mathematical Properties

Shannon entropy in statistical physics used to be introduced for a classical state $\rho(x)$ defined on continuous phase space $\{x\}$. For convenience of classical information theory, however, we shall use states $\rho(\mathrm{x})$ defined on discrete space $\{\mathrm{x}\}$, cf. Sect. 2.7.

Shannon entropy of a given state will be defined as follows:

$$S(\rho) = -\langle \log \rho \rangle = -\sum_{\mathrm{x}} \rho(\mathrm{x}) \log \rho(\mathrm{x}). \tag{9.1}$$

The Shannon entropy takes its minimum value 0 if the state is pure, say $\rho(\mathrm{x}) = \delta_{\mathrm{x}0}$. It takes its maximum value if the state is random: $\rho(\mathrm{x}) = \text{const}$.

Mixing increases the entropy. The entropy of a weighted mixture is greater than (or equal to) the weighted sum of the constituents' entropies (concavity):

$$S(w_1\rho_1 + w_2\rho_2) \ge w_1 S(\rho_1) + w_2 S(\rho_2)\ ; \quad w_1 + w_2 = 1. \tag{9.2}$$

Correlation decreases the entropy. The entropy of a bipartite composite state is less than (or equal to) the sum of the entropies of the two reduced states. The equality only holds when the two subsystems are uncorrelated (subadditivity):

L. Diósi, *A Short Course in Quantum Information Theory*,
Lecture Notes in Physics, 827, DOI: 10.1007/978-3-642-16117-9_9,

$$S(\rho_{AB}) \le S(\rho_A) + S(\rho_B). \tag{9.3}$$

Reduction decreases the entropy. The entropy of reduced states is less than (or equal to) the entropy of the composite state:

$$S(\rho_{AB}) \ge S(\rho_A),\ S(\rho_B). \tag{9.4}$$

The relative entropy can be defined for two states $\rho$, $\rho'$ that belong to the same system:

$$S(\rho\|\rho') = -S(\rho) - \sum_{\mathrm{x}} \rho(\mathrm{x}) \log \rho'(\mathrm{x}). \tag{9.5}$$

The relative entropy is zero if $\rho = \rho'$ and it is positive otherwise.

## 9.2 Messages

According to the abstraction of classical information, a finite sequence of *letters* taken from an *alphabet* constitutes a message. Messages can be used to store or communicate the encoded data. The simplest alphabet is binary, it has two letters usually associated with 0 and 1, respectively. The binary letters are also called bits. Binary messages are sequences of binary digits and can be thought of integers in binary representation. The unit of classical information measure is one *random* bit, or one bit in short (meaning always one random bit). The information of a random $n$-letter binary message is thus $n$ bits. If the alphabet is not binary the messages can still be faithfully translated into binary messages. One random letter of the $K$-letter alphabet can, on average, be converted into log $K$ random binary letters (bits), i.e., into a log $K$-letter long random binary message. A shorter binary message could not encode the original letter faithfully. Therefore we say that a random letter of a $K$-letter alphabet contains log $K$ bits information.

If the letter is not random it can be compressed faithfully into less than log $K$ random bits. The original message is redundant and the measure of information is less than $n\log K$ bits. In general, we define the information of a message by the length of the corresponding shortest faithfully compressed binary message. According to Shannon's theory, this shortest length is $nS(\rho)$ in the limit $n \to \infty$ [1]. We outline the proof.

## 9.3 Data Compression

Consider an $n$-letter long message of a $K$-letter alphabet $\{a, b, c, \ldots, z\}$:

$$\mathrm{x}_1\mathrm{x}_2, \ldots, \mathrm{x}_n, \tag{9.6}$$

and suppose that the letters independently follow the same a priori probability distribution:

$$\rho(a) = p_1, \quad \rho(b) = p_2, \quad \rho(c) = p_3, \ldots, \rho(z) = p_K. \tag{9.7}$$

Physically, each letter is represented by a discrete $K$-state classical system and the statistics of $n$-letter long messages is represented by the collective state

$$\rho(\mathrm{x}_1)\rho(\mathrm{x}_2), \ldots, \rho(\mathrm{x}_n) \equiv \rho^{\times n}(\mathrm{x}_1, \mathrm{x}_2, \ldots, \mathrm{x}_n). \tag{9.8}$$

The letters are random if $\rho = \text{const}$, otherwise the data contained in the above collective state can be compressed.

Indeed, many combinatorically possible messages are totally unlikely and become negligible if $n \to \infty$. Only a portion of the combinatorically possible messages will dominate statistically. These are called *typical messages*. In them, the relative frequency of each letter is identical to its a priori probability (9.7)! If in a very long message this rule were violated for any letter then the message would become atypical and thus statistically irrelevant. Therefore in typical messages the letters of the alphabet occur with frequencies

$$np_1, \; np_2, \ldots, np_K. \tag{9.9}$$

The typical messages can only differ from each other by the order of the letters! All typical messages have identical a priori probability. It is thus sufficient if we calculate the number of the typical messages, their probability distribution is even and equals the inverse of their number.

The number of the typical messages equals the number of polynomial combinations of $n$ elements satisfying the frequencies (9.9):

$$\text{No. of typical messages} = \frac{n!}{(np_1)!(np_2)!\ldots(np_K)!}. \tag{9.10}$$

If $n$ is large, Stirling approximation

$$\log n! = n \log n - n \log \mathrm{e} + O(\log n) \tag{9.11}$$

applies to all factorials, yielding

$$\log \frac{n!}{(np_1)!(np_2)!\ldots(np_K)!} \approx -n \sum_{k=1}^{K} p_k \log p_k. \tag{9.12}$$

The r.h.s. is just $n$-times the Shannon entropy of a single (non-random) letter. Accordingly, the number of typical messages reads

$$\text{No. of typical messages} = 2^{nS(\rho)}. \tag{9.13}$$

Their distribution is even. The collective state $\rho^{\times n}$ of the original messages can, information theoretically, be replaced by the even distribution of the typical messages:

$$\rho^{\times n} \longrightarrow \frac{1}{2^{nS(\rho)}}\Omega_n, \tag{9.14}$$

where $\Omega_n(\mathrm{x}_1, \mathrm{x}_2, \ldots, \mathrm{x}_n)$ is the indicator function which is 1 for the $2^{nS(\rho)}$ typical messages and zero otherwise. The typical ones of the original redundant messages (9.6) can therefore be numbered by $nS(\rho)$ binary digits constituting $nS(\rho)$-bit random messages. These binary messages are not redundant, cannot be further compressed faithfully. We have in such a way obtained the shortest faithful binary compression. Each letter of the original message has been faithfully compressed from $\log K$ bits into $S(\rho)$ bits.

This data compression theorem is the information theoretic interpretation of the Shannon entropy. The theorem as well as its interpretation are asymptotic, valid for one letter in infinite long redundant messages. This explains the success of the above heuristic proof. It must be obvious that for finite $n$ the frequencies of the letters would fluctuate around the asymptotic values (9.9), the set of typical messages might increase. Nevertheless, the exact proof would confirm that the above heuristic one has yielded the correct asymptotic number of the statistically relevant messages.

## 9.4 Mutual Information

Let us introduce the notation $X = \{\mathrm{x}, \rho(\mathrm{x})\}$ for a single-letter message where $\rho(\mathrm{x})$ is the a priori distribution of single letters. Shannon entropy $S(\rho)$ has been attributed to the distribution $\rho(\mathrm{x})$ of single letters. For terminological and notational conveniences we say that this is the entropy of the (single-letter) message $X$ and we introduce the alternative notation $H(X)$ for it:

$$H(X) = S(\rho). \tag{9.15}$$

The Shannon entropy measures the rate of our a priori ignorance concerning long messages. We must learn $H(X)$ bits per letter to eliminate our ignorance perfectly.

Suppose two separate messages $X$ and $Y$ are letter-wise correlated according to a certain distribution $\rho(\mathrm{x}, \mathrm{y})$. Let $H(X, Y)$ stand for the joint Shannon entropy. We define the mutual information of the two messages:

$$I(X\!:\!Y) = H(X) + H(Y) - H(X,Y) = \left\langle \log \frac{\rho(\mathrm{x},\mathrm{y})}{\rho(\mathrm{x})\rho(\mathrm{y})} \right\rangle. \tag{9.16}$$

This is zero if the two messages $X$ and $Y$ are uncorrelated. Otherwise it is positive. The mutual information $I(X\!:\!Y)$ gives the number of bits, per composite letter $(\mathrm{x}, \mathrm{y})$, that we can spare when we compress the joint message $(X, Y)$ instead of compressing $X$ and $Y$ separately.

The mutual information is related to the *conditional entropy*. While $I(X\!:\!Y)$ is symmetric function of the two messages the conditional entropy is not. The conditional entropy of the message $X$ is built upon the conditional probability $\rho(x|y) = \rho(x, y)/\rho(y)$:

$$H(X|Y) = -\langle \log \rho(x|y) \rangle = H(X, Y) - H(Y). \tag{9.17}$$

Conditional entropy means the minimum number of bits (per letter) needed to encode the message $X$ if the message $Y$ is perfectly known. $H(X|Y) = H(X)$ if the two messages are uncorrelated, and $H(X|Y) = 0$ if they are perfectly correlated. The mutual information and the two conditional entropies are simply related:

$$I(X{:}Y) = H(X) - H(X|Y) = H(Y) - H(Y|X). \tag{9.18}$$

## 9.5 Channel Capacity

Let $X$ be the input message of a noisy channel and let $Y$ be its output message. The channel is characterized by the *transfer probability* $\rho(x|y)$. Then the joint distribution of the input and output letters becomes

$$\rho(x, y) = \rho(y|x)\rho(x). \tag{9.19}$$

In fact, we can learn the output message $Y$ and, from it, we have to estimate the input message $X$. According to Bayes rule, we write down the aposteriori distribution of the input letter conditioned on the received output letter:

$$\rho(x|y) = \frac{\rho(x, y)}{\rho(y)} = \frac{\rho(y|x)\rho(x)}{\sum_{x'} \rho(y|x')\rho(x')}. \tag{9.20}$$

This becomes proportional to the transfer probability $\rho(x|y)$ of the channel if $\rho(x) = \text{const}$, i.e., when the input message is random:

$$\rho(x|y) = \frac{\rho(y|x)}{\sum_{x'} \rho(y|x')}. \tag{9.21}$$

The mutual information $I(X{:}Y)$ (9.18) indicates how many bits of the input message $X$ can faithfully be decoded from the perfect knowledge of one output letter. If the channel is noiseless, $\rho(x|y) = \delta_{xy}$, then $H(X) = H(Y) = H(X, Y) = I(X : Y)$ meaning that the output yields the input perfectly. In the contrary case, the channel noise makes the output message random: $\rho(y|x) = \text{const}$. Then $H(X) = H(X|Y)$ and $I(X{:}Y) = 0$ so we cannot estimate the input at all.

In the intermediate cases we obtain partial information on the input. Out of the $H(X)$ bits information per input letter, the output still contains $I(X{:}Y)$ bits information per letter. Those $I(X{:}Y)$ bits are in general not sufficient to faithfully estimate the input letter itself. Yet, the $I(X{:}Y)$ bits represent faithful partial information on the input message. The given channel communicates $I(X{:}Y)$ bits/

letter faithful information. This rate depends on the transfer probability $\rho(\mathrm{y}|\mathrm{x})$ of the channel and on the distribution $\rho(\mathrm{x})$ of the input letters, see also (9.18, 9.19). By an optimum choice of $\rho(\mathrm{x})$ we can maximize the the rate of faithful communication. This rate is called the channel capacity

$$C \equiv \max_{\{\rho(\mathrm{x})\}} I(X:Y), \tag{9.22}$$

which is the function of the transfer probability $\rho(\mathrm{y}|\mathrm{x})$ alone.

## 9.6 Optimal Codes

Shannon entropy $H(X)$ of a message $X$ has been interpreted as the bit-length of the shortest faithful code per letter. Our heuristic proof 9.3 could be made exact. First, one has to construct the block-code of $n$-letter long messages in such a way that, in the asymptotic limit $n \to \infty$ the code length is $\sim nH(X)$ bits. Note that the block-code cannot be decomposed into letter-wise codes. Second, one has to prove that this code is faithful, i.e., the decoding yields the original message at error rate which goes to zero for $n \to \infty$. Third, one has to prove that further data compression would necessarily violate faithfulness. Shannon has done the exact proof, that is his noiseless coding theorem.

In a similar way could we make the heuristic proof of channel capacity 9.5 exact. We consider $nC$-bit random messages to communicate through the channel faithfully. First we encode them properly into the $\sim n$-letter long input messages $X$ of the channel. Note that $C \leq H(X)$, i.e., the chosen code must be redundant if the channel is noisy. Second, we prove that, in the limit $n \to \infty$, decoding the output message $Y$ yields the initial $nC$ bits faithfully. Third, we prove that for the given channel any higher rate of faithful communication is impossible. This is Shannon's noisy code theorem.

## 9.7 Cryptography and Information Theory

The cryptography situation 6.4.2 can be interpreted in terms of classical information theory. There are three parties: Alice, Bob, and Eve. Alice generates the random key $X$, Bob estimates a certain key $Y$ which is a noisy version of $X$, while Eve steals a key $Z$ which is another noisy version of $X$. The quality of a given key-distribution protocol, whether classical or quantum, depends on the three mutual informations $I(X:Z)$, $I(Y:Z)$ and $I(X:Y)$. The success and faithfulness of key-distribution is measured by $I(X:Y)$, the loss of security is measured by $I(X:Z)$ and $I(Y:Z)$. In the ideal case $I(X:Y) = H(X) = H(Y)$ and $I(X:Z) = I(Y:Z) = 0$. The ultimate security of q-key protocols against unnoticed evedropping is based, among others, on the analysis of these mutual informations [2].

## 9.8 Entropically Irreversible Operations

We introduced the qualitative notion of irreversibility on examples, e.g., of the reduced dynamics $\mathcal{M}(t)$ in Sect. 2.5 where it meant just the non-invertibility of the operation $\mathcal{M}(t)$. The informatical concept of irreversibility is more restrictive than the concept of non-invertibility. It assumes information loss or, in alternative terminology: entropy increase. In typical reduced dynamics, e.g., the entropy for some initial states is increasing while for some other ones it is decreasing. Below, we are going to characterize the general class of entropy increasing operations.

A non-invertible operation $\mathcal{M}$ is (entropically) irreversible if it increases or preserves the entropy:

$$S(\mathcal{M}\rho) \geq S(\rho) , \tag{9.23}$$

for all states $\rho$. We shall show that a non-invertible $\mathcal{M}$ is irreversible if it leaves the totally mixed state $\rho_0(\mathrm{x}) = \mathrm{const}$ invariant: $\mathcal{M}\rho_0 = \rho_0$.

On a discrete phase space, the class of invertible operations is narrow: it corresponds to the permutations of the discrete points $\{\mathrm{x}\}$ of the state space. All other discrete classical operations $\mathcal{M}$ are non-invertible. We are going to show that $\mathcal{M}$ increases (or at least preserves) the entropy for all $\hat{\rho}$ if $\mathcal{M}\hat{\rho}_0 = \hat{\rho}_0$. The general form of an operation $\rho \to \rho' \equiv \mathcal{M}\rho$ reads: $\rho'(\mathrm{y}) = \sum_{\mathrm{x}} \rho(\mathrm{y}|\mathrm{x})\rho(\mathrm{x})$ where $\rho(\mathrm{y}|\mathrm{x})$ is the non-negative transfer function, satisfying $\sum_{\mathrm{y}} \rho(\mathrm{y}|\mathrm{x}) = 1$ for all x. From the condition $\mathcal{M}\rho_0 = \rho_0$ it follows that our transfer function is normalized in x as well: $\sum_{\mathrm{x}} \rho(\mathrm{y}|\mathrm{x}) = 1$ for all y. Such transfer function is also called doubly-stochastic. We can always write

$$S(\rho') - S(\rho) = \sum_{\mathrm{x,y}} \rho(\mathrm{x})\rho(\mathrm{y}|\mathrm{x}) \log \frac{\rho(\mathrm{x})}{\rho'(\mathrm{y})}. \tag{9.24}$$

We invoke the inequality $\ln \lambda > 1 - \lambda^{-1}$ valid for $\lambda \neq 1$ and apply it to $\lambda = \rho/\rho'$. This yields

$$S(\rho') - S(\rho) \geq \frac{1}{\ln 2} \sum_{\mathrm{x,y}} \rho(\mathrm{y}|\mathrm{x})[\rho(\mathrm{x}) - \rho'(\mathrm{y})] = 0. \tag{9.25}$$

In general, the strict inequality holds for $\rho \neq \rho_0$ unless the support of $\rho$ is smaller than the whole state space and the restriction of the non-invertible $\mathcal{M}$ for that support becomes invertible.

Let us see the simplest example. If the channel yields totally random output message always then it is entropically irreversible, it will increase the entropy for all possible messages apart from the marginal cases mentioned above. As a contrary example, consider a "broken" channel whose output message is a permanent 0 for any input whatsoever. The channel operation is obviously non-invertible but it is not entropically irreversible.

## 9.9 Problems, Exercises

9.1 *Positivity of relative entropy*. Using the inequality $\ln \lambda > 1 - \lambda^{-1}$, let us prove the positivity of the relative entropy.

9.2 *Concavity of entropy*. Why and how does mixing increase the entropy? Let us argue information theoretically. Method: start from two messages of different lengths proportional to $w_1$ and $w_2$, and of different single-letter distributions $\rho_1, \rho_2$, respectively.

9.3 *Subadditivity of entropy*. Using the positivity of the relative entropy, let us prove the subadditivity of the entropy.

9.4 *Coarse graining increases entropy*. Consider a given classical state, i.e. probability distribution $\rho(\mathrm{x})$ defined on the period [0,1] represented to a precision of $k$ binary digits: $\mathrm{x} \equiv 0.\mathrm{x}_1\mathrm{x}_2, \ldots, \mathrm{x}_k$. Let us introduce the coarse-grained state

$$\tilde{\rho}(\tilde{\mathrm{x}}) = \sum_{\mathrm{x}_k=0,1} \rho(\mathrm{x}),$$

defined on the same state space [0,1] to a precision of one digit less, i.e., we introduced $\tilde{\mathrm{x}} \equiv 0.\mathrm{x}_1\mathrm{x}_2, \ldots, \mathrm{x}_{k-1}$. Let us prove that coarse graining increases the entropy: $S(\tilde{\rho}) \geq S(\rho)$. Method: use the fact that state reduction increases the entropy.

## References

1. Shannon, C.E.: Bell Syst. Tech. J. **27**, 379–623 (1948)
2. Bennett, C.H.: Phys. Rev. Lett. **68**, 3121 (1992)

# Chapter 10
# Q-Information Theory

The q-information theory is similar to the classical one. The carrier of information is a q-system and this causes differences as well. A particular difference is that qubits carry information in double sense. First, we can encode classical bits into qubits. Second, unknown qubits carry information which is hidden and protected—as a consequence of the universal limitations of single q-state determination, cloning, or distinguishability.

## 10.1 Von Neumann Entropy, Mathematical Properties

Von Neumann entropy, or q-entropy, of a given q-state $\hat{\rho}$ will be defined as follows:

$$S(\hat{\rho}) = -\langle \log \hat{\rho} \rangle = -\mathrm{tr}(\hat{\rho} \log \hat{\rho}). \tag{10.1}$$

Obviously, the q-entropy is unitary invariant:

$$S(\hat{\rho}) = S(\hat{U}\hat{\rho}\hat{U}^\dagger). \tag{10.2}$$

The von Neumann entropy takes its minimum value zero if the state is pure, i.e.: $S(|\psi\rangle\langle\psi|) = 0$. It takes its maximum value on the maximally mixed state when $S(\hat{I}/d) = \log d$.

Let us construct the spectral expansion of the q-state $\hat{\rho}$:

$$\hat{\rho} = \sum_\lambda \rho_\lambda |\varphi_\lambda\rangle\langle\varphi_\lambda|. \tag{10.3}$$

The eigenvalues $\rho_\lambda$ of $\hat{\rho}$ form a probability distribution and its Shannon entropy equals the von Neumann entropy of the q-state $\hat{\rho}$:

L. Diósi, *A Short Course in Quantum Information Theory*,
Lecture Notes in Physics, 827, DOI: 10.1007/978-3-642-16117-9_10,

$$S(\rho) = -\sum_{\lambda} \rho_\lambda \log \rho_\lambda = S(\hat{\rho}). \tag{10.4}$$

Mixing increases the q-entropy. The q-entropy of a weighted mixture is greater than (or equal to) the weighted sum of the constituents' q-entropies (concavity):

$$S(w_1\hat{\rho}_1 + w_2\hat{\rho}_2) \geq w_1 S(\hat{\rho}_1) + w_2 S(\hat{\rho}_2); \quad w_1 + w_2 = 1. \tag{10.5}$$

Correlation decreases the q-entropy. The q-entropy of a bipartite composite q-system is less than (or equal to) the sum of the q-entropies of the two reduced q-states. The equality only holds when the two q-subsystems are uncorrelated (subadditivity):

$$S(\hat{\rho}_{AB}) \leq S(\hat{\rho}_A) + S(\hat{\rho}_B). \tag{10.6}$$

Reduction may change q-entropy in both directions, contrary to classical entropy. The difference between the q-entropies of the two reduced q-states cannot be greater than the entropy of the bipartite composite q-state (triangle inequality):

$$S(\hat{\rho}_{AB}) \geq |S(\hat{\rho}_A) - S(\hat{\rho}_B)|. \tag{10.7}$$

This is a genuine q-feature. The Shannon entropy of a *classical* sub-system is never greater than the entropy of the composite system (9.4):

$$S(\rho_{AB}) \geq S(\rho_A), S(\rho_B). \tag{10.8}$$

Hence the information content of a classical sub-system is smaller than (or equal to) the composite system's entropy. For q-systems this shall not be true. Typically, a pure entangled q-state has zero q-entropy while its subsystems are in mixed reduced states of positive q-entropy. Contrary to quantum, the classical pure composite states are so trivial that their subsystems remain pure.

The relative q-entropy can be defined for two states $\hat{\rho}, \hat{\rho}'$ that belong to the same q-system:

$$S(\hat{\rho}\|\hat{\rho}') = -S(\hat{\rho}) - \mathrm{tr}(\hat{\rho} \log \hat{\rho}'). \tag{10.9}$$

The relative entropy is zero if $\hat{\rho} = \hat{\rho}'$ and it is positive otherwise. The inequality $S(\hat{\rho}'\|\hat{\rho}) \geq 0$ is also called as the Klein inequality.

## 10.2 Messages

Consider a classical message $X = \{\mathrm{x}, \rho(x)\}$ which we encode letter-wise into certain, not necessarily orthogonal, pure states of a given d-state q-system:

$$\begin{array}{rcccc} \text{letter:} & \mathrm{x}_1 & \mathrm{x}_2 & \dots & \mathrm{x}_n \\ \text{q-code:} & |\mathrm{x}_1\rangle & |\mathrm{x}_2\rangle & \dots & |\mathrm{x}_n\rangle \\ \text{probability:} & \rho(\mathrm{x}_1) & \rho(\mathrm{x}_2) & \dots & \rho(\mathrm{x}_n) \end{array} \tag{10.10}$$

The sequence $|x_1\rangle, |x_2\rangle, \ldots, |x_n\rangle$ is a q-message of length $n$. Q-messages can be used to store or communicate the classical data encoded into q-states. We shall make an ultimate q-measurement on each q-code $|x\rangle$ in order to infer the encoded classical information to the best possible extent. A single pure state $|x\rangle$ can be unitarily transformed into $\log d$ qubits, and the above q-message can be transformed unitarily into a binary q-message of $n \log d$ qubits. The unit of q-information measure is one *random* qubit, or one qubit in short (meaning always one random qubit). The information of a random $n$-qubit q-message is thus $n$ qubits. We can decide if a q-message is random by constructing the density matrix of the 1-letter q-message:

$$\hat{\rho} = \sum_x \rho(x)|x\rangle\langle x|. \tag{10.11}$$

The q-message is random if this density matrix is totally mixed: $\hat{\rho} = \hat{I}/d$. Then the $n$-letter q-message is unitarily equivalent with $n \log d$ random qubits. If the q-message is not random it can be compressed faithfully into less than $n \log d$ random qubits. The original q-message is redundant and the measure of q-information is less than $n \log d$ qubits. In general, we define the information of a q-message by the length of the corresponding shortest faithfully compressed binary q-message. According to Schumacher's theory [1], this shortest length is $nS(\hat{\rho})$ in the limit $n \to \infty$. We outline the proof which resembles the proof of the classical case in Sect. 9.3.

## 10.3 Data Compression

From the viewpoint of q-theory, the q-message $|x_1\rangle, |x_2\rangle, \ldots, |x_n\rangle$ is a multiple composite pure state in a $d^n$-dimensional Hilbert space:

$$|\Psi\rangle = |x_1\rangle \otimes |x_2\rangle \otimes \cdots \otimes |x_n\rangle. \tag{10.12}$$

This state can be realized by $n$ independent d-state q-systems. The average of this state over all classical $n$-letter messages $x_1, x_2, \ldots, x_n$ will be the following collective (4.53) mixed state:

$$\sum_{\{x\}} \rho(x_1)\ldots\rho(x_n)|\Psi\rangle\langle\Psi| = \hat{\rho} \otimes \hat{\rho} \otimes \cdots \otimes \hat{\rho} \equiv \hat{\rho}^{\otimes n}. \tag{10.13}$$

If $\hat{\rho} \neq \hat{I}/d$, the q-message $|\Psi\rangle$ is not random; it is redundant and the measure of q-information is less than $n \log d$ qubits. Indeed, many directions $|\Psi\rangle$ in the Hilbert space are totally unlikely to occur and become negligible if $n \to \infty$. Only a subspace of the whole $d^n$-dimensional Hilbert space will dominate statistically. This subspace is called the *typical subspace*. We shall represent it by a hermitian projector $\hat{\Omega}_n$ which will be determined for asymptotically large $n$. It can be shown that the maximum faithful compression of the pure state messages $|\Psi\rangle$ corresponds to the maximum faithful compression of the *average* message $\hat{\rho}^{\otimes n}$. The task reduces to the maximum faithful compression of the collective state $\hat{\rho}^{\otimes n}$.

Of course, classical compression of the initial messages $\mathrm{x}_1, \mathrm{x}_2, \ldots, \mathrm{x}_n$ may not lead to an optimum compression of the encoded q-messages. Yet, as we shall argue, this becomes true provided the q-codes $|\mathrm{x}\rangle$ are orthogonal. We shall temporarily assume such orthogonal codes and utilize Shannon's classical data compression theorem. Note then that $S(\hat{\rho}) = S(\rho)$. Furthermore, we can perfectly discriminate the different q-messages (10.12) because they are orthogonal to each other so that we can simply use the corresponding projective q-measurement. Consequently, the best q-compression turns out to be equivalent with the best classical data compression learnt in Sect. 9.3. We only retain those q-messages (10.12) which correspond to the typical classical messages. We learned that their number is $2^{nS(\rho)}$. The corresponding composite vectors $|\Psi\rangle$ span a $2^{nS(\rho)} = 2^{nS(\hat{\rho})}$ - dimensional typical subspace. Its orthogonal complement in the $d^n$-dimensional Hilbert space is called the *atypical subspace*. Since the probability of each classical typical message is just $2^{-nS(\hat{\rho})}$, the faithful compression of the average message yields the following new density matrix:

$$\hat{\rho}^{\otimes n} \longrightarrow \frac{1}{2^{nS(\hat{\rho})}} \hat{\Omega}_n, \tag{10.14}$$

where $\hat{\Omega}_n$ is the hermitian projector onto the $2^{nS(\hat{\rho})}$-dimensional typical subspace. Q-data compression means that the original redundant q-messages (10.12) will be orthogonally projected by $\hat{\Omega}_n$ into the typical subspace. The typical subspace is unitary equivalent with $nS(\hat{\rho})$ qubits. The compressed state corresponds to $nS(\hat{\rho})$ - qubit random q-messages. These binary q-messages are not redundant, cannot be further compressed faithfully. We have in such a way obtained the shortest faithful binary compression. Each q-letter $|\mathrm{x}\rangle$ of the original q-message has been faithfully compressed from $\log d$ qubits into $S(\hat{\rho})$ qubits.

It is time to come back to the original task (10.10) and allow for non-orthogonal q-codes $|\mathrm{x}\rangle$ as well. It can be shown that the desired faithful q-data compression depends on the 1-letter density matrix $\hat{\rho}$ only. Therefore the faithful q-data compression takes always the form (10.14). The projector $\hat{\Omega}_n$ will be calculated as before: for the given $\hat{\rho}$, there can always be constructed a hypothetical q-message (10.10) with orthogonal q-codes. Construction of $\hat{\Omega}_n$ is thus straightforward if we know $\hat{\rho}$. The q-data compression will be realized by the projection $\hat{\Omega}_n |\Psi\rangle$ into the typical subspace of $\hat{\rho}^{\otimes n}$. The compressed q-code becomes a $2^{nS(\hat{\rho})}$-dimensional statevector and these statevectors occur at random within the typical subspace. Such statevectors are equivalently described by $nS(\hat{\rho})$ qubits which are also random. This way we have outlined the proof of Schumacher's q-data compression theorem: the classical message $\mathrm{x}_1, \mathrm{x}_2, \ldots,$ when encoded into the non-orthogonal pure states $|\mathrm{x}_1\rangle, |\mathrm{x}_2\rangle, \ldots,$ carries $S(\hat{\rho})$ qubit q-information per letter for asymptotically long messages, where $\hat{\rho}$ is the density matrix of the single letter q-message and $S(\hat{\rho})$ is its von Neumann entropy.

Von Neumann entropy $S(\hat{\rho})$ has thus been interpreted as the qubit-length per letter of the shortest faithful q-code of the message whose 1-letter average density matrix is $\hat{\rho}$. The above heuristic proof could be made exact. First, one has to construct the q-block-code of $n$-letter long messages in such a way that, in the asymptotic limit $n \to \infty$ the q-code length is $\sim nS(\hat{\rho})$ qubits. Note that the q-block-code cannot be decomposed into letter-wise codes. It must be a multiply entangled pure state of all $\sim nS(\hat{\rho})$ qubits. Second, one has to prove that the compression is faithful, i.e., it does not decrease the accessible information 10.4. Third, one has to prove that further data compression would necessarily violate faithfulness. Schumacher has done the exact proof, that is his noiseless q-coding theorem.

## 10.4 Accessible Q-Information

In the previous section, certain classical messages $X = \{\mathrm{x}, \rho(\mathrm{x})\}$ have been letter-wise encoded into pure q-states. More generally, we can encode letters into mixed q-states, too:

$$\begin{array}{rcccc} \text{letter:} & \mathrm{x}_1 & \mathrm{x}_2 & \ldots & \mathrm{x}_n \\ \text{q-code:} & \hat{\rho}_{\mathrm{x}_1} & \hat{\rho}_{\mathrm{x}_2} & \ldots & \hat{\rho}_{\mathrm{x}_n} \\ \text{probability:} & \rho(\mathrm{x}_1) & \rho(\mathrm{x}_2) & \ldots & \rho(\mathrm{x}_n) \end{array} \tag{10.15}$$

The density matrix of the averaged 1-letter code reads

$$\hat{\rho} = \sum_{\mathrm{x}} \rho(\mathrm{x}) \hat{\rho}_{\mathrm{x}}. \tag{10.16}$$

So far we have not discussed how much information is accessible regarding the original classical message $X$, by measuring the q-code. Now we apply a non-projective q-measurement 4.4.2 to each q-code $\hat{\rho}_{\mathrm{x}}$. Given the set of effects $\{\hat{\Pi}_{\mathrm{y}}\}$, the measurement outcome y will be considered a letter of another classical message $Y$ whereas the probabilities of letters y are conditioned on the original letter x:

$$\rho(\mathrm{y}|\mathrm{x}) = \mathrm{tr}\big(\hat{\Pi}_{\mathrm{y}} \hat{\rho}_{\mathrm{x}}\big). \tag{10.17}$$

Formally, this is the transfer function of a noisy classical channel. From the output message $Y$ we can letter-wise estimate the input message $X$. For statistics of long messages, the faithful average information carried by one letter y is called the *information gain* and measured by the mutual information 9.4:

$$I_{\mathrm{gain}} = I(X{:}Y) = H(X) - H(X|Y). \tag{10.18}$$

For the noisy classical channel itself, this was considered the communicated amount of information and we searched its maximum by optimizing the input message, recall Sect. 9.4. This time, however, we take both the input message and its q-code granted and search the maximum of $I(X{:}Y)$ by optimizing the q-effects $\{\hat{\Pi}_{\mathrm{y}}\}$ of our non-projective q-measurement:

$$A \equiv \max_{\{\hat{\Pi}_y\}} I(X:Y). \tag{10.19}$$

This $A$ is called the accessible information referred to the classical message $X$, from its q-codes (10.15). In the marginal case, the q-codes are orthogonal pure states. Then the optimum q-effects are orthogonal projectors and their projective q-measurement yields complete information on the original message: $A = H(X)$. The general case is much more difficult.

## 10.5 Entanglement: The Resource of Q-Communication

Q-communication between Alice and Bob needs a q-channel typically. Lacking a q-channel, Alice and Bob can cope with teleportation 7.3.2 which needs shared entangled states. One shared Bell state allows teleportation of one qubit, $n$ Bell states allow teleportation of $n$ qubits. We say: entanglement is the resource of q-communication. And vica versa, q-communication is a resource of shared entanglement. Alice and Bob can share $n$ Bell states if they can communicate $n$ qubits through their q-channel.

Accordingly, in the lack of q-communication the amount of (shared) entanglement is constant or at least not increasing, cf. Sect. 8.5. Alice and Bob can transform their entanglement into various forms but the amount of entanglement remains the same all the time. In q-information theory we consider the so-called LOCC situation where Alice and Bob can do any local operation on their respective local systems, and they can also communicate classically: but they cannot make q-communication. An important field of q-information theory is studying various q-state manipulations at LOCC constraints.

The prototype of the entangled state is a maximally entangled Bell state

$$|\Phi^+_{AB}\rangle = \frac{|0;A\rangle \otimes |0;B\rangle + |1;A\rangle \otimes |1;B\rangle}{\sqrt{2}} \equiv \frac{|00\rangle + |11\rangle}{\sqrt{2}}. \tag{10.20}$$

This is the unit of entanglement measure, too. For it: $E = 1$. We say that one qubit is Alice's local qubit, the other qubit is Bob's. When they share $k$ such Bell states we can say that they share entanglement $E = k$. If they share $k$ pairs of qubits that are only partially entangled, e.g.:

$$|\Psi_{AB}\rangle = \cos\theta\,|00\rangle + \sin\theta\,|11\rangle, \tag{10.21}$$

then the entanglement per pair is smaller then $E = 1$, Alice and Bob share less entanglement than $E = k$. We made a reasonable choice earlier as to the measure of the partial entanglement. Accordingly, we calculate the reduced states of Alice and/or Bob:

$$\hat{\rho}_A = \hat{\rho}_B = \cos^2\theta\,|0\rangle\langle 0| + \sin^2\theta\,|1\rangle\langle 1|. \tag{10.22}$$

The partially entangled state (10.21) will, by our heuristic definition (7.13), possess an entanglement measured by the von Neumann entropy of the above reduced states:

$$E = S = -(\cos^2\theta)\log(\cos^2\theta) - (\sin^2\theta)(\log\sin^2\theta). \tag{10.23}$$

This heuristic definition fits to the minimum ($E = 0$) and maximum ($E = 1$) entanglements. The interpretation of the intermediate values is only possible if we learn the reversible cycle of entanglement distillation–dilution. We are going to prove in the next section that, at LOCC constraints, a large number $n$ of *partially* entangled states $|\Psi_{AB}\rangle$ can be *distilled* into $k = Sn$ *maximally* entangled states $|\Phi^+_{AB}\rangle$. This LOCC operation is reversible: a large number $k$ of *maximally* entangled states $|\Phi^+_{AB}\rangle$ can be *diluted* into $k/S$ *partially* entangled states $|\Psi_{AB}\rangle$, as we see in the section after the next. Then it is crucial to understand that neither the efficiency of distillation nor the efficiency of dilution can be further improved. Why? Because we would then multiply the number of Bell states in a single distillation–dilution or dilution–distillation cycle. And this is impossible by LOCC, cf. Sect. 8.5.

In summary: the entanglement $E$ of a composite pure state is interpreted by the rate of the maximum number of distillable Bell states, and this, as we are going to prove, is indeed equivalent to the heuristically found definition $E = S$.

## 10.6 Entanglement Concentration (Distillation)

Suppose Alice and Bob share $n$ copies of partially entangled state $|\Psi_{AB}\rangle$ with entanglement $E = S$; let $n$ be large. We show that, using LOCC only, Alice and Bob can distill their states into $k \sim nE = nS$ maximally entangled Bell states:

$$|\Psi_{AB}\rangle^{\otimes n} \longrightarrow |\Phi^+_{AB}\rangle^{\otimes k}. \tag{10.24}$$

To this end, Alice performs a smart collective q-measurement on her $n$ qubits. The measured physical quantity will be the overall polarization in direction $z$. In the computational representation, this is equivalent with the collective measurement of the sum $\hat{m}$ of the binary physical quantities $\hat{x}$ (5.2) for Alice's $n$ qubits, where:

$$\begin{aligned}\hat{m} &= \hat{x}\otimes\hat{I}^{\otimes(n-1)} + \text{permutations}\\ &\equiv \hat{x}\otimes\hat{I}^{\otimes(n-1)} + \hat{I}\otimes\hat{x}\otimes\hat{I}^{\otimes(n-2)} + \cdots + \hat{I}^{\otimes(n-1)}\otimes\hat{x}.\end{aligned} \tag{10.25}$$

If the outcome is a given integer $m$ then the rules of projective q-measurement impose the following state change:

$$|\Psi_{AB}\rangle^{\otimes n} \longrightarrow \frac{|11\rangle^{\otimes m} \otimes |00\rangle^{\otimes(n-m)} + \text{permutations}}{\sqrt{\binom{n}{m}}}. \tag{10.26}$$

where the permutations include all states where $m$ qubits are $\mathrm{x} = 1$ and the other $n - m$ qubits are $\mathrm{x} = 0$. The most probable value of measurement outcome $m$ is approximately this:

$$n \sin^2 \theta, \tag{10.27}$$

and by that shall we approximate Alice's measurement outcome. If $n \to \infty$, the Stirling formula (9.11) applies to the average number $d$ of permutations on the r.h.s. of (10.26):

$$d = \binom{n}{m} \approx \binom{n}{n \sin^2 \theta} \approx 2^{nS} = 2^{nE}, \tag{10.28}$$

where $S$ just turns out to be the von Neumann entropy (10.23). It was adopted as the heuristic entanglement $E$ of the state $|\Psi_{AB}\rangle$. Hence Alice's collective measurement (10.26) has led approximately to the following entangled q-state:

$$\frac{|11\rangle^{\otimes m} \otimes |00\rangle^{\otimes(n-m)} + \text{permutations}}{\sqrt{d}}, \tag{10.29}$$

where the number $d$ of the mutually orthogonal terms also yields the dimension of the subspace spanned by themselves. The above expression is just the (7.3) Schmidt representation of the post-measurement state. The coefficients of the $d$ mutually orthogonal tensor product states are all equal to $1/\sqrt{d}$ and this state is therefore maximally entangled (7.11) on the $d$-dimensional subspace. Suppose log $d$ is integer, it can be shown this is of no ultimate restriction. Then, according to Sect. 7.1.6, the state (10.29) is equivalent with

$$k = \log d = nS \tag{10.30}$$

uncorrelated maximally entangled Bell states $|\Phi_{AB}^{+}\rangle$. This way we have shown that $n$ partially entangled qubit pairs each of entanglement $E$ can, via LOCC, be distilled into at least $k = nE$ maximally entangled pairs.

## 10.7 Entanglement Dilution

Suppose Alice and Bob share $k$ examples of the maximally entangled Bell states $|\Phi_{AB}^{+}\rangle$; let $k$ be large. We show that, using LOCC only, Alice and Bob can dilute their states into $n \sim k/E = k/S$ partially entangled states $|\Psi_{AB}\rangle$ of entanglement $E = S$:

$$|\Phi_{AB}^{+}\rangle^{\otimes k} \longrightarrow |\Psi_{AB}\rangle^{\otimes n}. \tag{10.31}$$

To this end, Alice and Bob will combine q-data compression and teleportation.

In addition to her $k$ qubits, Alice prepares $n = k/S = k/E$ local examples of the desired state $|\Psi_{AB}\rangle$ of partial entanglement $E = S$:

$$\left(\cos\theta\,|0;A'\rangle \otimes |0;A''\rangle + \sin\theta\,|1;A'\rangle \otimes |1;A''\rangle\right)^{\otimes n} \equiv |\Psi_{A'A''}\rangle^{\otimes n}. \tag{10.32}$$

In addition to his $k$ qubits, also Bob prepares $n = k/S = k/E$ local raw qubits in any states, say in a certain $\hat{\rho}_{B'}^{\otimes n}$. If Alice can, using LOCC only, transmit each qubits from her systems $A''$ to Bob's systems $B'$ then Alice and Bob would share $n$ example of the desired state $|\Psi_{A'B'}\rangle$ of partial entanglement $E = S$. Indeed, Alice can initiate teleportation which is typical LOCC operation. But Alice and Bob can teleport $k$ qubits only since they share just $k$ Bell pairs $|\Phi_{AB}^{+}\rangle^{\otimes k}$ initially. Fortunately, q-data compression will map the $n$ qubits into $k$.

Consider the collective state (10.32) and calculate the reduced collective state of the $n$ qubits on the systems $A''$:

$$\left(\cos^2\theta\,|0;A''\rangle\langle 0;A''| + \sin^2\theta\,|1;A''\rangle\langle 1;A''|\right)^{\otimes n} \equiv \hat{\rho}_{A''}^{\otimes n}. \tag{10.33}$$

This is the state Alice must teleport to Bob. Observe that $\hat{\rho}_A = \hat{\rho}_B = \hat{\rho}_{A'} = \hat{\rho}_{A''}$, cf. (10.22), therefore the corresponding von Neumann entropies $S$ coincide. Alice must compress $\hat{\rho}_{A''}^{\otimes n}$. We learned in Sect. 10.3 that for asymptotically large $n$ the collective state $\hat{\rho}_{A''}^{\otimes n}$ of $n$ qubits can faithfully be compressed into the $2^{nS} = 2^k$ dimensional subspace $\hat{\Omega}_n$ of just $k$ qubits. Therefore Alice teleports these $k$ qubits (the compressed q-codes) into the subspace $\hat{\Omega}_n$ of $k$ qubits of Bob's raw state $\hat{\rho}_{B'}^{\otimes n}$. For $n \to \infty$, the resulting state of the non-local composite systems $A'B'$ becomes the $n = k/S = k/E$ partially entangled states $|\Psi_{A'B'}\rangle$, identical to the desired ones $|\Psi_{AB}\rangle$.

## 10.8 Entropically Irreversible Operations

We introduced the qualitative notion of q-irreversibility on examples, e.g., of q-measurements in Sect. 4.4 and reduced dynamics in Sect. 4.5, where it meant just the non-invertibility of the corresponding q-operations. The informatical concept of q-irreversibility is more restrictive than the concept of non-invertibility. It assumes q-information loss or, in alternative terminology: q-entropy increase. In non-selective q-measurements the von Neumann entropy is always increased or preserved. In typical reduced dynamics, however, the q-entropy for some initial q-states is increasing while for some other ones it is decreasing. Below, we are going to characterize the general class of entropy increasing q-operations.

A non-invertible (non-unitary) q-operation $\mathcal{M}$ is (entropically) irreversible if it increases or preserves the von Neumann entropy:

$$S(\mathcal{M}\hat{\rho}) \geq S(\hat{\rho}), \tag{10.34}$$

for all q-states $\hat{\rho}$. We shall show that a non-invertible $\mathcal{M}$ is irreversible if it leaves the totally mixed q-state $\hat{\rho}_0 = \hat{I}/d$ invariant: $\mathcal{M}\hat{\rho}_0 = \hat{\rho}_0$.

The q-entropy of a state $\hat{\rho}$ is always equal to the Shannon entropy computed from the eigenvalues $\{w_\lambda\}$, provided we take the degenerate eigenvalues twice or more, accordingly. Similarly, the q-entropy of $\hat{\rho}' \equiv \mathcal{M}\hat{\rho}$ coincides with the Shannon entropy computed from the eigenvalues $\{w'_\mu\}$ of $\hat{\rho}'$. Therefore it is sufficient if we prove the classical relationship $S(w') \geq S(w)$ where $w, w'$ are the diagonal (eigenvalue) distributions of the density matrices $\hat{\rho}, \hat{\rho}'$, respectively. Let us introduce the spectral expansions $\hat{\rho} = \sum_\lambda w_\lambda \hat{P}_\lambda$ and $\hat{\rho}' = \sum_\mu w'_\mu \hat{P}'_\mu$. Then we express the eigenvalues of $\hat{\rho}'$ through the following equivalent expressions:

$$w'_\mu = \mathrm{tr}\left(\hat{P}'_\mu \hat{\rho}'\right) = \mathrm{tr}\left(\hat{P}'_\mu \mathcal{M}\hat{\rho}\right) = \sum_\lambda \mathrm{tr}\left(\hat{P}'_\mu \mathcal{M}\hat{P}_\lambda\right) w_\lambda, \tag{10.35}$$

where the last form has been obained by substituting the spectral expansion of $\hat{\rho}$. For *fixed* $w, w'$, the above map looks like a classical operation on $w$, yielding $w'$ through a transfer function. A transfer function is always normalized in the first variable $\mu$. Our transfer function is normalized in the second variable as well:

$$\sum_\lambda \mathrm{tr}\left(\hat{P}'_\mu \mathcal{M}\hat{P}_\lambda\right) = \mathrm{tr}\left(\hat{P}'_\mu \mathcal{M}\hat{I}\right) = 1, \tag{10.36}$$

since $\mathcal{M}\hat{I} = \hat{I}$ according to our initial postulation. Our transfer function is double stochastic, therefore the Shannon entropy of $w$ will be greater or equal to the Shannon entropy of $w'$, cf. Sect. 9.8. In general, the strict inequality holds for $\hat{\rho} \neq \hat{\rho}_0$ unless $\hat{\rho}$ is singular and the restriction of the non-unitary $\mathcal{M}$ for the support of $\hat{\rho}$ becomes invertible.

The simplest example of (entropically) irreversible q-operations is the non-selective q-measurement. It is non-invertible and it preserve the totally mixed state. As a contrary example, consider a "broken" q-channel whose output message is a permanent $|0\rangle$ for any input whatsoever: $\hat{\rho}' = \hat{P}_0\hat{\rho}\hat{P}_0 + \hat{\sigma}_x\hat{P}_1\hat{\rho}\hat{P}_1\hat{\sigma}_x$ where $\hat{P}_\mathrm{x} = |\mathrm{x}\rangle\langle\mathrm{x}|$, for $\mathrm{x} = 0, 1$. The q-channel operation is obviously non-invertible but it is not entropically irreversible.

## 10.9 Problems, Exercises

10.1 *Subadditivity of q-entropy*. Using the Klein inequality, let us prove the subadditivity of the von Neumann entropy.

10.2 *Concavity of q-entropy, Holevo entropy.* When we prepare a q-state as a mixture $\hat{\rho} = \sum_n w_n \hat{\rho}_n$ then the entropy of the mixture is never smaller than the average entropy of the components:

$$0 \leq S(\hat{\rho}) - \sum_n w_n S(\hat{\rho}_n).$$

The r.h.s. is called Holevo entropy. It puts an upper bound on the accessible information and it is bounded from above by the so-called entropy of preparation $S(w)$:

$$A \leq S(\hat{\rho}) - \sum_n w_n S(\hat{\rho}_n) \leq S(w).$$

The first inequality is called the Holevo bound [2] and it is a most powerful theorem of q-information theory.
Holevo entropy is non-negative because of the concavity of the q-entropy. Let us prove concavity from subadditivity.

10.3 *Data compression of the non-orthogonal code.* Suppose random bits are encoded into non-orthogonal pure states exactly like in the case of the q-banknote 6.4.1 as well as of q-cryptography 6.4.2. Thus we consider the states $|\uparrow z\rangle, |\uparrow x\rangle$ the q-code of the random classical binary message $X = \{\mathrm{x}, \rho(\mathrm{x})\}$:

$$\begin{array}{lcc} \text{x letter:} & 0 & 1 \\ \text{q-code:} & |\uparrow z\rangle & |\uparrow x\rangle \\ \rho(\mathrm{x}) \text{ probability:} & 1/2 & 1/2 \end{array} \tag{10.37}$$

The Shannon entropy of the random classical letter is $H(X) = 1$ bit. Let us determine the maximum faithful q-data compression rate.

10.4 *Distinguishing two non-orthogonal qubits: various aspects.* We have previously mentioned two strategies 6.3 to distinguish non-orthogonal random states like $|\uparrow z\rangle, |\uparrow x\rangle$. The first strategy 6.3.1 used a single projective measurement of either $\hat{\sigma}_z$ or $\hat{\sigma}_x$ chosen at random. It was perfectly conclusive at rate 25%. The second strategy 6.3.2 used a single non-projective measurement regarding three suitable q-effects $\hat{\Pi}_z, \hat{\Pi}_x, \hat{\Pi}_?$, and this strategy was perfectly conclusive at rate $1 - 1/\sqrt{2} \approx 30\%$ and perfectly inconclusive in the rest. In the context of Prob. 10.3, the information gain $I_{gain}$ will coincide with the rate 30% of the unambiguous decisions. Yet, there are different projective q-measurements which are never perfectly conclusive but their information gain is bigger. Let us calculate $I_{gain}$ for the single polarization measurement in the direction $(1, 0, \ -1)$.

10.5 *Simple optimum q-code.* Assume we have to encode the three colors R,G, and B, occuring at random, into the respective pure states $|R\rangle, |G\rangle, |B\rangle$ of a qubit.

Suppose we do not care if the q-code would not guarantee the highest accessible information. Suppose we have limited q-channel capacities therefore we want optimum q-code in a sense that it cannot be shortened further, i.e., without the loss of the accessible information. Then we use our q-channel economically. Let us construct the corresponding three pure states $|R\rangle$, $|G\rangle$, and $|B\rangle$.

## References

1. Schumacher, B.: Phys. Rev. A **51**, 2738 (1995)
2. Holevo, A.S.: Problems Inf. Transm. **5**, 247 (1979)

# Chapter 11
# Q-Computation

Classical digital computer performs its algorithm via one- and two-bit operations realized by one- and two-bit "gates". The (yet hypothetical) q-digital computer would do the same. The classical logical gates become q-gates, their "circuits" are called q-circuits. They can perform what is called q-parallel computation. Q-algorithms are always reversible. Therefore we learn classical reversible computation first then we discuss the earliest q-algorithms that might overcome all classical algorithms targeting a similar task.[1]

## 11.1 Parallel Q-Computing

A classical computer works like this: the input *storage* stores an $n$-digit binary number x upon which the *algorithm* is performed and the result will be read out from the same storage. Now, let the computer be a q-system, operating coherently on its q-storage which is an $n$-qubit state vector

$$|x_1\rangle \otimes \ldots \otimes |x_n\rangle \equiv |x\rangle; \quad x \equiv x_1x_2\ldots x_n. \tag{11.1}$$

Unprecedented in classical computation, we can store all possible $N = 2^n$ initial values in the q-storage if we *superpose* them, say, with the same amplitudes:

$$|S\rangle = \frac{1}{\sqrt{N}}\sum_{x=0}^{N-1}|x\rangle. \tag{11.2}$$

We call this state the totally superposed state of the storage. What about a coherent q-algorithm? It is a unitary $N \times N$ matrix $\hat{U}$. Given a classical algorithm, we would naively think that its q-counterpart corresponds to a certain matrix $\hat{U}$. This

[1] For a short review on q-computation, see Ekert et al. [1].

L. Diósi, *A Short Course in Quantum Information Theory*,
Lecture Notes in Physics, 827, DOI: 10.1007/978-3-642-16117-9_11,

is the case, indeed, provided the classical algorithm is reversible. If it is not then we have to replace it by a reversible version. Fortunately, it had been known in classical theory of computation that any computational task can be realized reversibly if some redundant data are drawn along, cf. Sect. 11.2. This way we can realize any classical computational task via the corresponding unitary transformation $\hat{U}$ on a q-storage. Parallel computation becomes possible on all possible initial data captured by the totally superposed state $|S\rangle$. Moreover, we could perform various different reversible algorithms running simultaneously on the same q-computer.

Contrary to early expectations [2], it is not so easy to find a concrete computational task whose coherent q-algorithm outperforms all possible incoherent classical algorithms. The total superposition of the initial states $|S\rangle$ is instrumental but, surprisingly, it does not in itself guarantee an advantage. We need to invent genuine q-algorithms, like those of Sects. 11.3 and 11.4, differing radically from the classical ones.

## 11.2 Evaluation of Arithmetic Functions

We are going to discuss a basic classical computational task. The algorithm should compute functions $y = f(x)$ that map $n$-bit integers x onto $m$-bit integers y:

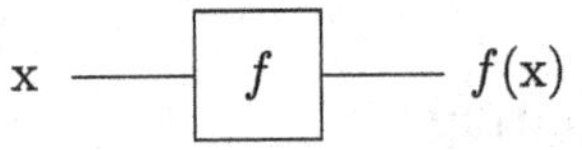

$$x \quad \boxed{\quad f \quad} \quad f(x) \tag{11.3}$$

The map $f$ is not necessarily invertible, therefore the corresponding algorithm is not necessarily reversible. For the purpose of q-parallel computation, we extend the map $f$ in such a way that the new map $F$ be reversible even if $f$ were not. Accordingly, we set up independent input and output storages, i.e., the output storage exists already at input time. Let the new invertible map $F$ be the following:

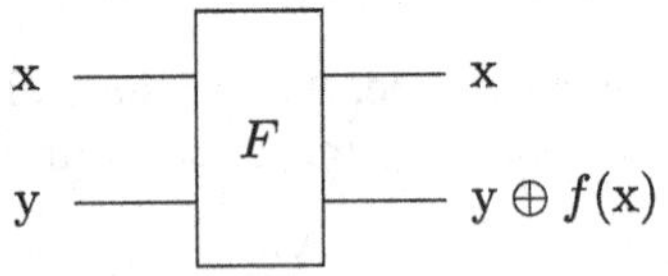

$$
\begin{array}{l}
x \quad \underline{\qquad} \boxed{\quad F \quad} \underline{\qquad} \quad x \\
y \quad \underline{\qquad} \phantom{\boxed{\quad F \quad}} \underline{\qquad} \quad y \oplus f(x)
\end{array}
\tag{11.4}
$$

The summation is understood mod $2^m$. How does this algorithm calculate $f(x)$? That's simple. We set the output storage at initial time to zero, $y = 0$, so that the same storage at final time will contain $f(x)$.

The new function $F(x, y)$ is always invertible. Therefore the computation of the arithmetic function $f(x)$ has been realized by a reversible classical algorithm which can directly be converted into the following q-algorithm:

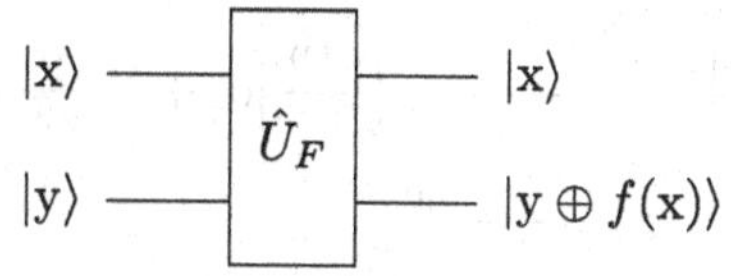


(11.5)

Instead of this diagrammatic definition, the $2^{n+m} \times 2^{n+m}$ unitary matrix of the q-algorithm can be defined in the usual way as well:

$$\hat{U}_F(|\mathrm{x}\rangle \otimes |\mathrm{y}\rangle) = |\mathrm{x}\rangle \otimes |\mathrm{y} \oplus f(\mathrm{x})\rangle. \tag{11.6}$$

If we set the input storage to the totally superposed state $|S\rangle$ instead of $|\mathrm{x}\rangle$ and we set the output storage to $|0\rangle$ then the following state emerges:

$$\hat{U}_F(|S\rangle \otimes |0\rangle) = \frac{1}{\sqrt{N}} \sum_{\mathrm{x}=0}^{N-1} |\mathrm{x}\rangle \otimes |f(\mathrm{x})\rangle. \tag{11.7}$$

In a single unitary step one has thus evaluated the arithmetic function $f(\mathrm{x})$ for all possible arguments $\mathrm{x} = 0, 1, \ldots, N-1$. However, we can read out just one value when we measure the final states of both the input and output registers. The advantage of parallel q-computation is related to the usage of the state $|S\rangle$ but requires more than that.

## 11.3 Oracle Problem: The First Q-Algorithm

Consider a black-box (11.3) to evaluate a certain arithmetic function $f(\mathrm{x})$, and call it the (classical) oracle. If we ask the oracle a "question" x, it provides the immediate "answer" $\mathrm{y} = f(\mathrm{x})$. This time it is irrelevant how it evaluates the function. Suppose the point is that we do not know the function $f(\mathrm{x})$, only the oracle knows it. The hypothetical task will be this: we want to learn the function $f(\mathrm{x})$ therefore we are going to ask questions and evaluating the answers. How many questions should we ask?

It is this class of problems where the first ever task has been invented [3, 4] for which a smart q-algorithm is clearly more efficient than any classical algorithm. In the simplest case, let $f$ map a single bit x into a single bit y. Suppose we have to learn whether $f(\mathrm{x})$ is a constant function or it is not. If $f(\mathrm{x})$ is evaluated by a classical oracle then we must ask twice: we submit both initial values $\mathrm{x} = 0$ and $\mathrm{x} = 1$ in separate "questions". More questions do not exist, less questions are not sufficient to decide whether or not $f(0) = f(1)$. If, however, the function $f$ is evaluated by a q-oracle (11.5) then a single "question" will be sufficient:

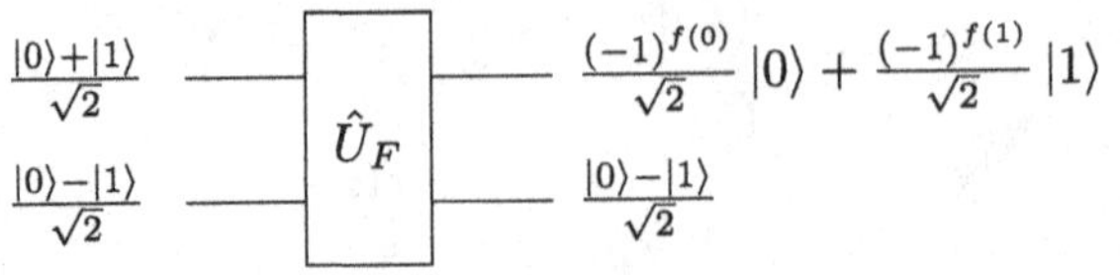
$\frac{|0\rangle+|1\rangle}{\sqrt{2}}$
$\frac{|0\rangle-|1\rangle}{\sqrt{2}}$
$\hat{U}_F$
$\frac{(-1)^{f(0)}}{\sqrt{2}}|0\rangle + \frac{(-1)^{f(1)}}{\sqrt{2}}|1\rangle$
$\frac{|0\rangle-|1\rangle}{\sqrt{2}}$

(11.8)

The input storage should be set to the total superposition $|S\rangle$ of all possible states (i.e.: $\mathrm{x} = 0,1$) in accordance with the concept of q-parallel computation (11.2). We do not set the output storage to zero, rather we set it to the superposition of the two possible states $\mathrm{x} = 0,1$—with the opposite phases. We show that to such a single "question" the q-oracle's answer is decisive regarding the constancy of $f$. As we see from (11.8), the output storage never changes. The input storage, however, does:

$$\begin{array}{ll} \frac{|0\rangle + |1\rangle}{\sqrt{2}} & \text{if } f(0) = f(1)\,, \\ \frac{|0\rangle - |1\rangle}{\sqrt{2}} & \text{if } f(0) \neq f(1)\,. \end{array} \tag{11.9}$$

These two input storage states are orthogonal. A single ultimate projective q-measurement can discriminate them. After the measurement we know with certainty whether the function $f$ is constant or not.

We have yet to prove that the q-oracle provides the "answer" (11.8) for the smart "question" there. Obviously we can re-write the initial state this way:

$$\frac{|0\rangle + |1\rangle}{\sqrt{2}} \otimes \frac{|0\rangle - |1\rangle}{\sqrt{2}} = \frac{1}{\sqrt{2}} \sum_{\mathrm{x}=0,1} |\mathrm{x}\rangle \otimes \frac{|0\rangle - |1\rangle}{\sqrt{2}}\,. \tag{11.10}$$

The unitary map $\hat{U}_F$ (11.6) applies to this, and yields the following state:

$$\frac{1}{\sqrt{2}} \sum_{\mathrm{x}=0,1} |\mathrm{x}\rangle \otimes \frac{|f(\mathrm{x})\rangle - |1 \oplus f(\mathrm{x})\rangle}{\sqrt{2}}\,. \tag{11.11}$$

We can write the state of the output storage into an equivalent form

$$|f(\mathrm{x})\rangle - |1 \oplus f(\mathrm{x})\rangle = (-1)^{f(\mathrm{x})}(|0\rangle - |1\rangle)\,. \tag{11.12}$$

Indeed, the state of the output storage has, upto its phase, remained unchanged. Using the above form, the final composite state produced by the q-oracle becomes

$$\frac{1}{\sqrt{2}} \sum_{\mathrm{x}=0,1} (-1)^{f(\mathrm{x})} |\mathrm{x}\rangle \otimes \frac{|0\rangle - |1\rangle}{\sqrt{2}}\,. \tag{11.13}$$

It coincides with the "answer" displayed on the diagram (11.8).

The advantage of q-tests over classical ones becomes more spectacular when the oracle $f$ maps a large number of $n$ bits into one. Suppose the arithmetic function $f$ is either constant or balanced. This latter means that $f(\mathrm{x})$ returns 0 and 1 equal times when x runs over the $N = 2^n$ different values. Suppose that we are interested to learn whether the oracle is constant or balanced. Classically we must ask $N/2 + 1$ "questions". In case of the corresponding q-oracle, a single "question" is enough. The construction is very similar to the former simplest one. We keep setting the input storage to the totally superposed state $|S\rangle$ and we keep setting the output storage to $(|0\rangle - |1\rangle)/\sqrt{2}$. We analyze the "answer" of the oracle by an ultimate projective q-measurement of $|S\rangle\langle S|$. We are going to prove that $f$ is constant if the outcome is 1 and $f$ is balanced if the outcome is 0. The proof consists of the same steps as before, with the only change that x runs from 0 to $N-1$. Accordingly, the initial state reads

$$|S\rangle \otimes \frac{|0\rangle - |1\rangle}{\sqrt{2}} = \frac{1}{\sqrt{N}} \sum_{\mathrm{x}=0}^{N-1} |\mathrm{x}\rangle \otimes \frac{|0\rangle - |1\rangle}{\sqrt{2}}, \tag{11.14}$$

while the final state, i.e., the "answer" will be

$$\frac{1}{\sqrt{N}} \sum_{\mathrm{x}=0}^{N-1} (-1)^{f(\mathrm{x})} |\mathrm{x}\rangle \otimes \frac{|0\rangle - |1\rangle}{\sqrt{2}}. \tag{11.15}$$

Obviously, the final state of the input storage is $|S\rangle$ when $f$ is constant and it is orthogonal to $|S\rangle$ when $f$ is balanced.

## 11.4 Searching Q-Algorithm

Suppose that the function $f(\mathrm{x})$, mapping $n$ bits into one bit, is always zero except for $\mathrm{x} = \mathrm{x}_0$:

$$f(\mathrm{x}) = \delta_{\mathrm{x}\mathrm{x}_0}. \tag{11.16}$$

Let a classical oracle evaluate the function. Assume that $\mathrm{x}_0$ is unknown to us and we have to learn it by asking the oracle. This is the basic searching problem formulated in the language of the oracle-problem. The unknown integer $\mathrm{x}_0$ can take any values between 0 and $N-1$ where $N = 2^n$. Obviously, we must ask $N-1$ questions in the worst case, and $O(N)$ questions in general . Grover showed that we can ask $O(\sqrt{N})$ questions of the corresponding q-oracle, which is much less than $O(N)$. The corresponding q-algorithm [5] will be much faster than any classical algorithm to solve the same searching problem.

Grover's q-algorithm is iterative, of $R \sim \sqrt{N}$ iterations. In each cycle, we ask the q-oracle one question and perform a given unitary transformation on its answer. This state becomes the question in the forthcoming cycle, etc. The output storage of the oracle stays always in the state $(|0\rangle - |1\rangle)/\sqrt{2}$, hence the "question" $|\mathrm{x}\rangle$ submitted into the input storage will produce this answer:

$$(-1)^{f(\mathrm{x})}|\mathrm{x}\rangle = (1 - 2\delta_{\mathrm{x}\mathrm{x}_0})|\mathrm{x}\rangle. \tag{11.17}$$

This "question-to-answer" map can be described by the following $N \times N$ unitary matrix:

$$\hat{I} - 2|\mathrm{x}_0\rangle\langle\mathrm{x}_0|. \tag{11.18}$$

Thus we use the q-oracle to mark for us the searched direction $|\mathrm{x}_0\rangle$ in the state-space. Let our first "question" be the totally superposed state $|S\rangle$. The "answer" will be $(\hat{I} - 2|\mathrm{x}_0\rangle\langle\mathrm{x}_0|)\,|S\rangle$ and, according to Grover, we transform it by another unitary matrix

$$\hat{I} - 2|S\rangle\langle S|. \tag{11.19}$$

The result will be utilized as the "question" asked of the q-oracle in the second iteration cycle, etc. We are going to prove that after $R$ iterations the resulting state is approximately $|\mathrm{x}_0\rangle$:

$$\left[(\hat{I} - 2|S\rangle\langle S|)(\hat{I} - 2|\mathrm{x}_0\rangle\langle\mathrm{x}_0|)\right]^R |S\rangle \approx |\mathrm{x}_0\rangle\ , \tag{11.20}$$

so that the projective q-measurement of $\hat{\mathrm{x}}$ on the resulting state yields the searched value $\mathrm{x}_0$.

The proof is elementary. Why? The iteration (11.20) does nothing else than a rotation of the initial vector $|S\rangle$ toward the searched vector $|\mathrm{x}_0\rangle$, by the same angle in each cycle. The process can be described in the real two-dimensional plane spanned by $|S\rangle$ and $|\mathrm{x}_0\rangle$ . In this plane, rotations (11.18, 11.19) are considered modulo $\pi$ since the sign of the state vector is irrelevant. For simplicity's sake, we assume $N \gg 1$. Then $|\mathrm{x}_0\rangle$ is almost orthogonal to $|S\rangle$ :

$$\langle\mathrm{x}_0|S\rangle = \frac{1}{\sqrt{N}}\ . \tag{11.21}$$

Their angle $(\pi/2) - \epsilon$ is almost the rectangle. The defect is very small:

$$\epsilon \approx \frac{1}{\sqrt{N}}. \tag{11.22}$$

In the first iteration, the matrix (11.18) reflects the initial state $|S\rangle$ w.r.t. the direction $|\mathrm{x}_0\rangle$; the resulting vector will be reflected by the matrix (11.19) w.r.t. the direction of $|S\rangle$. The outcome vector is such that it encloses a smaller angle $(\pi/2) - 3\epsilon$ with the searched vector $|\mathrm{x}_0\rangle$ . It is easy to see that the two reflections in each further cycle of Grover iteration will always rotate toward $|\mathrm{x}_0\rangle$ by the same angle $2\epsilon$. Since the initial vector $|S\rangle$ was away from $|\mathrm{x}_0\rangle$ by an angle $(\pi/2) - \epsilon$, the necessary number of iterations will be

$$R \approx \frac{\pi}{4\epsilon} \approx \frac{\pi}{4}\sqrt{N}\ . \tag{11.23}$$

The $R$-times iterated state approaches $|x_0\rangle$ to within a small angle $\sim\epsilon$. Hence its projective measurement in the computational basis yields $x_0$ with an error probability as little as $\sim 1/N$.

## 11.5 Fourier Algorithm

The well-known discrete classical Fourier transformation is equivalent with a unitary transformation on the given complex vector space. We can use the q-physics notations. Let $\{|x\rangle\}$ stand for the basis vectors where x runs through the $N$ binary numbers of $n = \log N$ bits. The classical Fourier transformation is given by

$$\hat{U}\,|x\rangle = \frac{1}{\sqrt{N}}\sum_{y=0}^{N-1} e^{2\pi ixy/N}|y\rangle. \tag{11.24}$$

Equivalently, the elements of the $N \times N$ unitary Fourier matrix read

$$\langle y|\hat{U}|x\rangle = \frac{1}{\sqrt{N}}e^{2\pi ixy/N}. \tag{11.25}$$

Since this classical map is reversible we can directly turn it into a q-algorithm. Yet, as we shall see, the entanglement of $\hat{U}|x\rangle$ is only apparent, it is a multiple direct product state in fact. This factorization makes the classical fast-Fourier algorithm possible.

We are going to prove that, using the notation $x = x_1x_2\ldots x_n$, the Fourier transform is a product state of $n$ different qubits. We substitute the digital expansion $y = N\sum_{k=1}^{n} y_k/2^k$ in order to factorize the elements of the unitary matrix of the Fourier transform:

$$e^{2\pi ixy/N} = \prod_{k=1}^{n}\exp(2\pi ixy_k/2^k) = e^{2\pi iy_1 0.x_n}e^{2\pi iy_2 0.x_{n-1}x_n}\ldots e^{2\pi iy_n 0.x_1x_2\ldots x_n}. \tag{11.26}$$

We can also factorize the sum over the computational basis states:

$$\sum_{y} e^{2\pi i\ldots}|y\rangle = \sum_{y_1=0}^{1}\sum_{y_2=0}^{1}\cdots\sum_{y_n=0}^{1} e^{2\pi i\ldots}|y_1\rangle\otimes|y_2\rangle\otimes\cdots\otimes|y_n\rangle. \tag{11.27}$$

Substituting these expressions into that of the Fourier transform (11.24), we obtain the following:

$$\hat{U}\,|x\rangle = \frac{|0\rangle + e^{2\pi iy_1 0.x_n}|1\rangle}{\sqrt{2}}\otimes\frac{|0\rangle + e^{2\pi iy_2 0.x_{n-1}x_n}|1\rangle}{\sqrt{2}}\otimes\ldots\otimes\frac{|0\rangle + e^{2\pi iy_n 0.x_1x_2\ldots x_n}|1\rangle}{\sqrt{2}} \tag{11.28}$$

This shows that the q-algorithm of the Fourier transform can be built up from single qubit operations controlled by the state of other qubits. Thus, in fact, these operations are multi-qubit operations. Although the number of control qubits is more than one, two-qubit controlled gates (cf. Sect. 11.8) are universally sufficient when organized in suitable q-circuits, cf. Fig. 11.2.

## 11.6 Period Finding Q-Algorithm

Let $f(\mathrm{x})$ be an arithmetic function that maps $n = \log N$ bit integers into $m$ bit integers. Suppose it is periodic with period $r$, i.e., for all $0 \leq \mathrm{x} \leq N - r - 1$:

$$f(\mathrm{x} + r) = f(\mathrm{x}) \,. \tag{11.29}$$

Suppose that $r$ is unknown to us and we have to determine it. According to the widespread classical method, we calculate the Fourier transform of $f(\mathrm{x})$:

$$\tilde{f}(\mathrm{y}) = \frac{1}{\sqrt{N}} \sum_{\mathrm{x}=0}^{N-1} \mathrm{e}^{2\pi \mathrm{ixy}/N} f(\mathrm{x}) \,, \tag{11.30}$$

which will show peaks at the multiples of the wave number $N/r$. The proof is the following. We consider the general case when $r$ does not divide $N$ while we assume that the integer number $\nu = [N/r]$ of complete periods is sufficiently big. If we approximate $\sum_{\mathrm{x}=0}^{N-1}$ by $\sum_{\mathrm{x}=0}^{\nu r-1}$ we obtain

$$\tilde{f}(\mathrm{y}) \sim \sum_{\mathrm{x}=0}^{\nu r-1} \mathrm{e}^{2\pi \mathrm{ixy}/N} f(\mathrm{x}) = \left( \sum_{\ell=0}^{\nu-1} \mathrm{e}^{2\pi \mathrm{i}\ell r\mathrm{y}/N} \right) \sum_{\mathrm{x}=0}^{r-1} \mathrm{e}^{2\pi \mathrm{ixy}/N} f(\mathrm{x}). \tag{11.31}$$

The pre-factor takes the closed form

$$\sum_{\ell=0}^{\nu-1} \mathrm{e}^{2\pi \mathrm{i}\ell r\mathrm{y}/N} = \frac{1 - \mathrm{e}^{2\pi \mathrm{i}\nu r\mathrm{y}/N}}{1 - \mathrm{e}^{2\pi \mathrm{i}r\mathrm{y}/N}} \,, \tag{11.32}$$

which has peaks if y is close to a multiple of $N/r$.

Similar is the Fourier q-algorithm of period finding. We evaluate the function $f(\mathrm{x})$ for all arguments $\mathrm{x} = 0, 1, \ldots, N - 1$ in a single unitary operation (11.7), i.e., we prepare the following state:

$$\frac{1}{\sqrt{N}} \sum_{\mathrm{x}=0}^{N-1} |\mathrm{x}\rangle \otimes |f(\mathrm{x})\rangle \,, \tag{11.33}$$

but we do not yet read it out. Rather we apply the Fourier transformation (11.24) to the input register, yielding

$$\frac{1}{\sqrt{N}}\sum_{x=0}^{N-1}\hat{U}|x\rangle\otimes|f(x)\rangle=\frac{1}{N}\sum_{y=0}^{N-1}|y\rangle\otimes\sum_{x=0}^{N-1}e^{2\pi ixy/N}|f(x)\rangle. \tag{11.34}$$

We measure the computational basis in the input state. The random measurement outcome y will have the following probability:

$$p(y)=\left\|\frac{1}{N}\sum_{x=0}^{N-1}e^{2\pi ixy/N}|f(x)\rangle\right\|^2. \tag{11.35}$$

This probability distribution is peaked if y is close to a multiple of $N/r$. The proof copies the classical proof: the approximation and reorganization of the sum in (11.30) can invariably be applied to the sum on the r.h.s. above. Finally we get the (squared modulus of the) classical pre-factor (11.32) of $\tilde{f}(y)$ as the pre-factor of $p(y)$:

$$p(y)\sim\left|\frac{1-e^{2\pi ivry/N}}{1-e^{2\pi iry/N}}\right|^2\times\left\|\sum_{x=0}^{r-1}e^{2\pi ixy/N}|f(x)\rangle\right\|^2. \tag{11.36}$$

This probability is enhanced at the multiples of $N/r$. Hence the outcome y is very likely to be the multiple of $N/r$, i.e., $y/N\approx k/r$ where both integers $k$ and $r$ are unknown. To find $r$ we look for the best rational number approximation to $y/N$ where the divisor is less (in our case: much less) than $N$, and we identify $r$ by that divisor. If a check (11.29) fails, we repeat the whole algorithm.

The period finding q-algorithm becomes instrumental for the Shor algorithm [6] to factorize large numbers, which is the most promising prediction of q-computation theory. The Shor q-algorithm is exponentially faster than all known classical algorithms to factorize large integers. This speed-up of the Shor algorithm is due to the q-algorithm of period finding as part of the Shor algorithm.

## 11.7 Error Correction

Uncontrollable environmental interactions, i.e., noise may corrupt the storage of computers. Error detection and correction of a classical storage are possible if we apply redundant codes. If we encode one logical bit into three identical raw bits then all bit-flip errors can be detected and corrected by "majority voting" provided the noise corrupts a single raw bit at a time. The corrupted q-storage is a little more difficult task because error detection and correction must respect and restore the faultless superpositions and entanglements. We must detect and correct the error of a qubit whereas the qubit itself remains unknown all the time.

Consider the error detection and correction of a single qubit $|x\rangle$. First we discuss how to protect the qubit $c_0|0\rangle+c_1|1\rangle$ against a bit-flip error. The simplest idea is that we encode our logical qubit into three raw qubits:

$$|\psi\rangle = c_0\,|000\rangle + c_1\,|111\rangle, \tag{11.37}$$

which occupy the two-dimensional code space

$$\hat{C} = |000\rangle\langle 000| + |111\rangle\langle 111| \tag{11.38}$$

within the eight-dimensional raw space. Suppose the first qubit gets corrupted by a bit-flip:

$$|\psi\rangle \rightarrow |\psi'\rangle = \hat{X}_1|\psi\rangle = c_0|100\rangle + c_1|011\rangle\,. \tag{11.39}$$

Observe that the unitary bit-flip maps the code space $\hat{C}$ onto the subspace $\hat{P}_1 = \hat{X}_1\hat{C}\hat{X}_1$ which is orthogonal to $\hat{C}$. Alternatively, if the bit-flip error corrupted the second or third qubit, the code space would map to $\hat{P}_2 = \hat{X}_2\hat{C}\hat{X}_2$ or $\hat{P}_3 = \hat{X}_3\hat{C}\hat{X}_3$, respectively. The projectors $\hat{P}_1, \hat{P}_2, \hat{P}_3$—called error syndromes—form an orthogonal set. If we measure them on the state $|\psi'\rangle$, we can detect the bit-flip error, and we can correct it. If the $k$'th qubit ($k = 1, 2, 3$) has been corrupted then the measurement outcome is $\hat{P}_k = 1$, and we know from it that we have to re-flip the $k$'th qubit:

$$|\psi'\rangle \rightarrow \hat{X}_k|\psi'\rangle = |\psi\rangle. \tag{11.40}$$

The unknown original state (11.37) recovers provided the bit-flip error has influenced just one of the three raw qubits.

A similar protocol applies if, instead of the bit-flip errors $\hat{X}_1, \hat{X}_2, \hat{X}_3$, we assume another unitary equivalent set of single-qubit errors like, e.g., the phase-flip errors $\hat{Z}_1, \hat{Z}_2, \hat{Z}_3$. Since $\hat{Z}_k = \hat{H}_k\hat{X}_k\hat{H}_k$ for $k = 1, 2, 3$, we must apply the unitary transform $\hat{H}_1\hat{H}_2\hat{H}_3$ to the basis vectors $\{|000\rangle, |111\rangle\}$ as well as to the error syndromes $\hat{P}_k$ of the bit-flip error correction protocol, this way we obtain the phase-flip error correction protocol.

The real issue is whether we can construct error correction protocols for all single qubit errors $\hat{X}, \hat{Y}, \hat{Z}$ together. Three raw qubits are not sufficient because the eight-dimensional Hilbert space can not host more than three error syndromes. But nine redundant raw qubits are already sufficient [7]:

$$|\psi\rangle = c_0\left(\frac{|000\rangle + |111\rangle}{\sqrt{2}}\right)^{\otimes 3} + c_1\left(\frac{|000\rangle - |111\rangle}{\sqrt{2}}\right)^{\otimes 3}. \tag{11.41}$$

Then the code space reads

$$\hat{C} = \left(\frac{|000\rangle + |111\rangle}{\sqrt{2}}\,\frac{\langle 000| + \langle 111|}{\sqrt{2}}\right)^{\otimes 3} + \left(\frac{|000\rangle - |111\rangle}{\sqrt{2}}\,\frac{\langle 000| - \langle 111|}{\sqrt{2}}\right)^{\otimes 3}. \tag{11.42}$$

There can be $9 \times 3 = 27$ different single-qubit errors:

$$|\psi\rangle \rightarrow |\psi'\rangle = (\hat{\sigma}_a)_k|\psi\rangle \quad (a = x, y, z; k = 1, 2, \ldots, 9) \tag{11.43}$$

where, for notational convenience, we returned to Pauli's symbols instead of $\hat{X}, \hat{Y}, \hat{Z}$. The error syndromes corresponding to the 27 errors are $\hat{P}_{ak} = (\hat{\sigma}_a)_k \hat{C} (\hat{\sigma}_a)_k$. One can directly inspect that these 27 projectors and $\hat{C}$ are all mutually orthogonal, the proof needs just patience. Now we possess an error correcting protocol where nine raw qubits encode one logical qubit, and we can correct any of the three unitary errors $\hat{\sigma}_x, \hat{\sigma}_y, \hat{\sigma}_z$ corrupting any single qubit. We measure the 27 projectors, if $\hat{P}_{ak} = 1$ for a given pair $a$, $k$ then we know that the $k$'th qubit has been corrupted by the unitary operation $\hat{\sigma}_a$. Let us apply the same unitary transformation $\hat{\sigma}_a$ once more to the $k$'th qubit of the state $|\psi'\rangle$, it recovers the correct state $|\psi\rangle$ (11.41).

However, the environmental influence on a single qubit can be much more complex than (11.43). It is, in general, the statistical mixture of linear maps (cf. Sect. 8.2), the resulting (unnormalized) state of a linear map can be written as

$$|\psi\rangle \rightarrow |\psi'\rangle = (\hat{I}_k + v_x(\hat{\sigma}_x)_k + v_y(\hat{\sigma}_y)_k + v_z(\hat{\sigma}_z)_k)|\psi\rangle, \tag{11.44}$$

where $v_x$, $v_y$, $v_z$ are complex numbers. There is a continuum of different errors! The ultimate good news is this: the nine-qubit protocol with its 27 different error syndromes will correct *all* single qubit errors. Indeed, the measurement of the 27 error syndromes first collapses the corrupted state $|\psi'\rangle$ into one of its four terms above. No correction is needed if all syndromes yield zero. In the contrary case, the discrete error correction recovers the same correct state $|\psi\rangle$ (11.41).

## 11.8 Q-Gates, Q-Circuits

Any given classical algorithm can be decomposed into a series of Boolean operations. It is known that a small selection of one- and two-bit operations suffice although this selection is not unique. The selected ones are called universal logical operations (or gates, referring to their technical realization). To perform a given algorithm, the classical logical gates are organized into logical circuits. Similarly, we define q-logical gates and circuits to perform q-algorithms. We have already learned about the 1-qubit logical operations (Sect. 6.1.1) and now we associate the notion of q-gate to them. We also mentioned that the Hadamard gate (*H*) and phase-gate (*T*), i.e.:

$$\hat{H} = \frac{1}{\sqrt{2}}\begin{bmatrix} 1 & 1 \\ 1 & -1 \end{bmatrix} \qquad |\mathrm{x}\rangle \;-\!\boxed{H}\!-\; \frac{(-1)^{\mathrm{x}}}{\sqrt{2}}|\mathrm{x}\rangle + \frac{1}{\sqrt{2}}|1-\mathrm{x}\rangle \tag{11.45}$$

$$\hat{T}(\varphi) = \begin{bmatrix} \mathrm{e}^{-\mathrm{i}\varphi/2} & 0 \\ 0 & \mathrm{e}^{\mathrm{i}\varphi/2} \end{bmatrix} \qquad |\mathrm{x}\rangle \;-\!\boxed{\varphi}\!-\; \exp\left[\mathrm{i}(\mathrm{x}-\tfrac{1}{2})\varphi\right]|\mathrm{x}\rangle \tag{11.46}$$

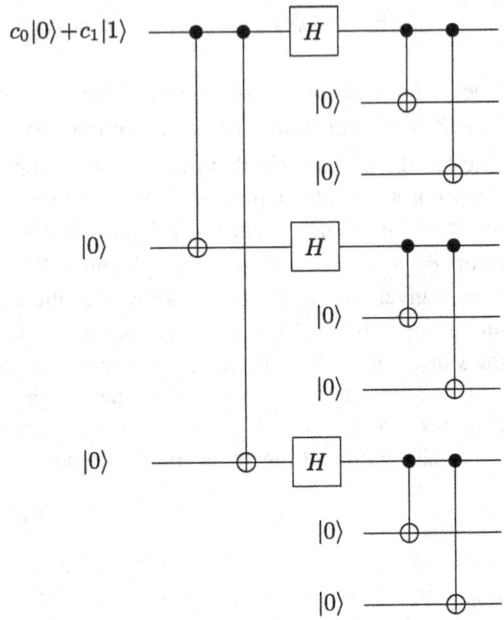


**Fig. 11.1** Q-circuit encoding Shor error correction. This structure (cf. [9] by Nielsen and Chuang), can easily be compared with the desired code state (11.41)

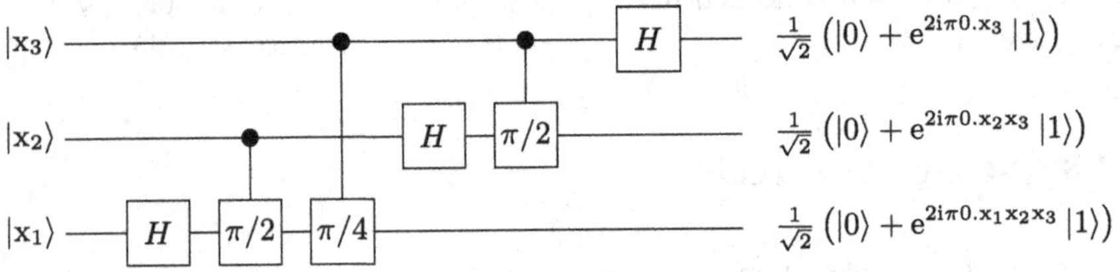


**Fig. 11.2** Q-circuit of the Fourier transform. This three-qubit ($n = 3$) example illustrates the general structure, cf. [9] by Nielsen and Chuang. An overall phase, not shown here, w.r.t. the expression (11.28) comes from the conventional choice (11.46) of the phase-gates

are universal one-qubit gates: all one-qubit unitary operations can be realized by them.

The two-qubit operations (gates) can be derived from the so-called controlled operations (gates). Then the first qubit is the control-qubit and the second is the target-qubit. If the control-qubit is 1 then a given unitary operation $\hat{U}$ is performed on the target-qubit. If the control qubit is 0 then nothing will happen:

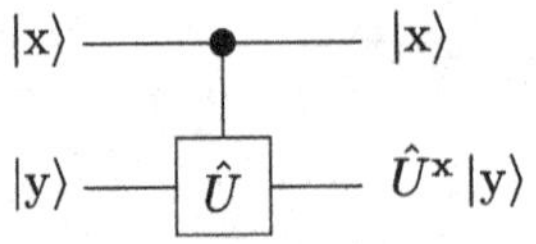


(11.47)

This is, for instance, the controlled NOT (cNOT, XOR):

$$\hat{C} = \begin{bmatrix} \hat{I} & 0 \\ 0 & \hat{X} \end{bmatrix} \equiv \begin{bmatrix} 1&0&0&0 \\ 0&1&0&0 \\ 0&0&0&1 \\ 0&0&1&0 \end{bmatrix} \tag{11.48}$$

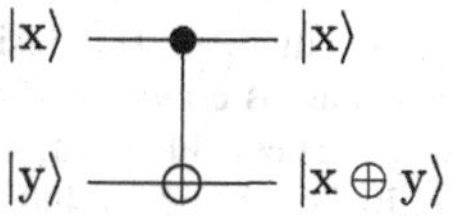


The $4 \times 4$ matrix corresponds to the order $|00\rangle, |01\rangle, |10\rangle, |11\rangle$ in obvious notation $|\mathrm{xy}\rangle = |\mathrm{x}\rangle \otimes |\mathrm{y}\rangle$ of the bipartite basis-vectors. The chosen symbol $\oplus$ on the diagram reminds us that in computational basis the cNOT sets the target-qubit to the modulo 2 sum of the two initial qubits.

It is known that the Hadamard-, the phase-, and the cNOT-gates are universal q-gates, cf. Problem 6.1. These three q-gates allow constructing all unitary operations on a q-storage of any number $n$ of qubits. Design of q-logical circuits, similarly to design of classical logical circuits, has its own standards and tricks [8]. Talking about q-algorithms, standard diagrams of q-circuits are used extensively. Moreover, the language and formalism of q-information theory is penetrating our understanding—and teaching—the traditional q-theory as well.

In order to illustrate the art of q-circuit design, we include two examples which give a first hint and further motivations to the interested.

## 11.9 Problems, Exercises

11.1 *Creating the totally symmetric state.* Let us prove that $n$ independent qubits, in state $|0\rangle$ each, can be transformed into the totally symmetric state $|S\rangle$ by using $n$ Hadamard gates.

11.2 *Constructing Z-gate from X-gate.* Let us confirm that the one-qubit q-gates $X$ and $Z$ are related via Hadamard gate sandwiches: $HXH = Z$ and $HZH = X$.

11.3 *Constructing controlled Z-gate from cNOT-gate.* Adding one-qubit gates to a cNOT-gate in a suitable way, let us build a q-circuit to realize a controlled-Z operation.

11.4 *Constructing controlled phase-gate from two cNOT-gates.* Adding one-qubit gates to two cNOT-gates in a suitable way, let us build a q-circuit to realize a controlled-phase operation.

11.5 *Q-circuit to produce Bell states.* Let us show that the following q-circuit produces the four Bell states from the computational basis:

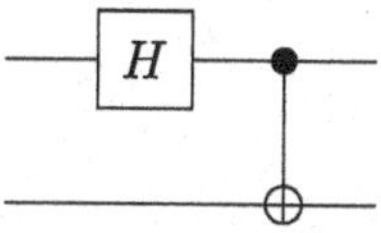


11.6 *Q-circuit to measure Bell states.* Let us construct the q-circuit of projective measurement in the Bell basis.

## References

1. Ekert, A., Hayden, P., Inamori, H.: Basic concepts in quantum computation (Les Houches lectures 2000); Los Alamos e-print arXiv: quant-ph/0011013
2. Feynman, R.P.: Int. J. Theor. Phys. **21**, 467 (1982)
3. Deutsch, D.: Proc. R. Soc. Lond. A **400**, 97 (1985)
4. Deutsch, D., Jozsa, R.: Proc. R. Soc. Lond. A **439**, 533 (1992)
5. Grover, L.K.: A fast quantum mechanical algorithm for database search. In: Proceedings of the 28th Annual STOC, Association for Computer Machinery, New York (1996)
6. Shor, P.W.: Algorithms for quantum computation: discrete logarithm and factoring. In: 35th Annual Symposium on Foundations of Computer Science. IEEE Press, Los Alamitos (1994)
7. Shor, P.W.: Phys. Rev. A **52**, 2493 (1995)
8. Tucci, R.R.: QC Paulinesia. http://www.ar-tiste.com/PaulinesiaVer1.pdf (2004)
9. Nielsen, M.A., Chuang, I.L.: Quantum Computation and Quantum Information. Cambridge University Press, Cambridge (2000)

# Chapter 12
# Qubit Thermodynamics

Entropy had been rooted in thermodynamics. We have to explore this background of the microscopic q-entropy. We do not hesitate to apply thermodynamics to a single qubit if we recall the classical ideal gas where thermodynamical equations are satisfied by the microscopic state of each single molecule. Since thermal environment is natural for a physical qubit, further notions of thermodynamics, like thermalization, refrigeration, or the Carnot cycle, apply to qubits. Their significance for q-information is a delicate challenge.

## 12.1 Thermal Qubit

According to the general principles of thermodynamics, a system placed into a thermal reservoir of temperature $T$ will reach thermal equilibrium with the reservoir. This principle is valid for a quantum system as well: it will under the influence of the reservoir relax to a well-defined thermal equilibrium state, the so-called Gibbs state $\mathcal{N}\exp(-\hat{H}/k_BT)$ ($k_B$ is the Boltzmann constant) which is obviously a stationary state of the von Neumann equation (4.5). When the quantum system is a single qubit, we can write the Hamiltonian matrix (5.24) in the form $\hat{H} = -\frac{1}{2}\epsilon\hat{\sigma}_z$. This may correspond to an electronic spin in vertical magnetic field where $\epsilon$ is the energy difference between the states $|\uparrow\rangle = |0\rangle$ and $|\downarrow\rangle = |1\rangle$. The same Hamiltonian matrix may also refer to a two-level atom where we identify its ground and excited states as $|0\rangle$ and $|1\rangle$, respectively. The Gibbs state of the qubit takes the form

$$\hat{\rho}_\beta = \frac{1}{2\cosh(\beta\epsilon/2)}\mathrm{e}^{\beta\epsilon\hat{\sigma}_z/2} = \frac{1}{1+\exp(-\beta\epsilon)}\left(|0\rangle\langle 0| + \mathrm{e}^{-\beta\epsilon}|1\rangle\langle 1|\right). \tag{12.1}$$

L. Diósi, *A Short Course in Quantum Information Theory*,
Lecture Notes in Physics, 827, DOI: 10.1007/978-3-642-16117-9_12,

We introduced the inverse temperature $\beta = 1/k_B T$. Recall the Fock representation from Sect. 5.5 where we talk about the occupation number $\hat{n} = |1\rangle\langle 1| = \hat{a}^\dagger \hat{a}$ of the excited state which in $\hat{\rho}_\beta$ yields

$$n_\beta = \mathrm{tr}(\hat{n}\hat{\rho}_\beta) = \frac{1}{1+\exp(\beta\epsilon)}. \tag{12.2}$$

For thermodynamic purposes, it is more convenient if we shift the Hamiltonian $\hat{H}$ by a constant so that $\hat{H} = \epsilon\hat{n}$, i.e., the ground state energy is zero. Let us determine the average energy of our thermal qubit:

$$E = \mathrm{tr}(\hat{H}\hat{\rho}_\beta) = \frac{\epsilon}{1+\mathrm{e}^{\beta\epsilon}}, \quad 0 \leq E \leq \frac{1}{2}\epsilon. \tag{12.3}$$

This we shall identify as the thermodynamic energy of the thermal qubit. We also calculate the von Neumann entropy of the Gibbs state:

$$S(\hat{\rho}_\beta) = -\,\mathrm{tr}(\hat{\rho}_\beta \log \hat{\rho}_\beta) = \frac{\log(1+\mathrm{e}^{\beta\epsilon})}{1+\mathrm{e}^{\beta\epsilon}} + \frac{\log(1+\mathrm{e}^{-\beta\epsilon})}{1+\mathrm{e}^{-\beta\epsilon}}. \tag{12.4}$$

With one eye on thermodynamics, we express the r.h.s. in terms of the energy $E$:

$$S(E) = -\frac{\epsilon - E}{\epsilon}\log\frac{\epsilon - E}{\epsilon} - \frac{E}{\epsilon}\log\frac{E}{\epsilon}, \quad 0 \leq E \leq \frac{1}{2}, epsilon. \tag{12.5}$$

This entropy $S(E)$ is our candidate thermodynamic function for the thermal qubit. We must give it the right physical dimension by a factor $k_B$, and we have to rescale it by a factor $\ln 2$ to turn the binary logarithm into the natural one:

$$S_{\mathrm{th}}(E) = (k_B \ln 2) S(E) = -k_B \frac{\epsilon - E}{\epsilon}\ln\frac{\epsilon - E}{\epsilon} - k_B\frac{E}{\epsilon}\ln\frac{E}{\epsilon}. \tag{12.6}$$

The entropy of a single thermal qubit satisfies the thermodynamics equation

$$\frac{\mathrm{d}S_{\mathrm{th}}(E)}{\mathrm{dE}} = \frac{1}{T}. \tag{12.7}$$

## 12.2 Ideal Qubit Gas

If we want to construct a bulk macroscopic thermodynamic system yet with the exactly tractable quantum dynamics, our simplest system is the abstract ideal gas of qubits. Suppose the gas consists of $N$ non-interacting distinguishable qubits as molecules, each having the same Hamiltonian $\hat{H} = \epsilon\hat{n}$. The thermal equilibrium state of the gas is

$$\hat{\rho}_\beta \otimes \hat{\rho}_\beta \otimes \cdots \otimes \hat{\rho}_\beta = \hat{\rho}_\beta^{\otimes N}, \tag{12.8}$$

where $\hat{\rho}_\beta$ is given by the Eq. (12.1). The average energy $E$ and the von Neumann entropy $S$ of the gas become $N$ times the single qubit average energy (12.3) and entropy (12.5), respectively. Therefore we obtain the thermodynamic entropy of the gas if we take $N$-times the single qubit entropy (12.6) whereas we replace the single qubit thermodynamic energy by $E/N$:

$$S_{\text{th}}(E,N) = N(k_B \ln 2)S(E/N). \tag{12.9}$$

This entropy is an extensive thermodynamic variable itself since it is a homogeneous linear function of the two extensive thermodynamic variables $E$ and $N$, as it should be. The entropy of the ideal qubit gas by construction satisfies the same thermodynamics equation (12.7) as its molecules do:

$$\frac{\partial S_{\text{th}}(E,N)}{\partial E} = \frac{1}{T}. \tag{12.10}$$

Another standard relationship $\partial S_{\text{th}}(E,N)/\partial N = -\mu/T$ defines the chemical potential $\mu$ of the qubit ideal gas.

The qubit ideal gas has exact additive thermodynamics in the number $N$ of the molecules. This is why single qubit thermodynamics ($N = 1$) introduced in the preceding section makes sense though it is hardly conform with the macroscopic notion of a thermodynamic system which would even suppose the thermodynamic limit $N \to \infty$.

## 12.3 Informatic and Thermodynamic Entropies

For a single qubit as well as for the qubit ideal gas, we derived the simple relationship between the thermodynamic and informatic (von Neumann) entropies:

$$S_{\text{th}} = (k_B \ln 2)S. \tag{12.11}$$

It turns out that this relationship is valid for general q-systems as well. However, we should ask how it is possible? The interpretation of the informatic entropy $S$ refers to the maximum compressibility of q-data, it tells us the length of the shortest faithful code that *we* can reach to compress our data, cf. Chaps. 9 and 10. Apparently, $S$ constrains *our* success to economize our resources like storage or channel capacities. On the other hand, the thermodynamic entropy $S_{\text{th}}$ constrains the objective physics of the given system. What has $S_{\text{th}}$ to do with $S$ then? Obviously nothing, unless "data compression" is present in the physical world independently of human intentions. In fact, this is the case.

We know from statistical physics that the thermal equilibrium of a system corresponds to a q-state $\hat{\rho}$ which is maximally randomized at the given conditions. By maximally randomized, statistical physics understands the maximum of the von Neumann entropy. One can directly prove that the Gibbs state $\hat{\rho}_\beta = \mathcal{N}\exp(-\beta\hat{H})$ of a q-system corresponds to the maximum of the von Neumann entropy $S(\hat{\rho})$ at fixed average energy $E = \mathrm{tr}(\hat{H}\hat{\rho})$. This fact is independent of the informatic context of the von Neumann entropy. But we can turn it around and give it the informatic interpretation. The equilibrium q-state $\hat{\rho}_\beta$ looks like a shortest code at the constraint of fixed average energy. This shows us that maximum "data compression" is a physical phenomenon. What was the redundant code, or what was the data? Such relevant further issues, ignored this time, might lead us to deeper insight into the information aspect of statistical physics and vice versa.

The equivalence (12.11) of thermodynamic and informatic entropies is established for all systems in thermal equilibrium. A rigorous proof is possible for homogeneous systems in the limit of infinite size. Whether the equivalence can be extended for non-equilibrium states is a principal issue. The concrete proof is, however, problematic because the full microscopic description of non-equilibrium states is technically difficult if not completely hopeless.

## 12.4 Q-Thermalization

As we said in Sect. 12.1, a certain qubit placed into a thermal reservoir of temperature $T$ will reach the thermal equilibrium state $\hat{\rho}_\beta$ (12.1). We are going to construct a simple model[1] of this thermalization, and we continue to prefer the Fock representation, cf. Sect. 5.5.

Our thermal reservoir will be the abstract qubit gas (Sect. 12.2) in thermal equilibrium. We consider a further "central" qubit in arbitrary initial state $\hat{\rho}$. The same Hamiltonian $\hat{H} = \epsilon\hat{n} = \epsilon\hat{a}^\dagger\hat{a}$ will be assumed for the reservoir qubits as well as for the central qubit. The central qubit is the one to be thermalized, i.e., led to the state $\hat{\rho}_\beta$ by the reservoir. The thermalization will be modelled through independent two-qubit interactions (collisions) between a reservoir qubit $\hat{\rho}_\beta$ and the central qubit. Let the interaction correspond to an instantaneous unitary transformation $\hat{U}$ which occurs at frequency $\nu$. If the reservoir is large, it is plausible to assume that no reservoir qubit will interact twice or more, i.e., the central qubit will always interact with an unperturbed reservoir qubit $\hat{\rho}_\beta$.

Consider a single collision $\hat{\rho}_\beta \otimes \hat{\rho} \rightarrow \hat{U}\hat{\rho}_\beta \otimes \hat{\rho}\hat{U}^\dagger$ between a certain reservoir qubit and the central qubit, described by a certain unitary $4 \times 4$ matrix $\hat{U}$. The resulting irreversible q-operation on the reduced density matrix of the central qubit reads

$$\hat{\rho} \longrightarrow \mathrm{tr}_\beta\left[\hat{U}\hat{\rho}_\beta \otimes \hat{\rho}\hat{U}^\dagger\right], \qquad (12.12)$$

[1] After [1] by Scarani et al.

where we refer to the partial trace over the reservoir qubit. Our choice for the unitary matrix $\hat{U}$ is this:

$$\hat{U} = \exp\left(g\hat{a}^\dagger \otimes \hat{a} - \text{h.c.}\right), \tag{12.13}$$

which is the general form to exchange the excitation energy between the reservoir and the central qubits, respectively. If we substitute $\hat{U}$ into (12.12), we can easily calculate the r.h.s. to the second order in the coupling $g$:

$$\hat{\rho} \longrightarrow \hat{\rho} + g^2(1-n_\beta)\left(\hat{a}\hat{\rho}\hat{a}^\dagger - \frac{1}{2}\{\hat{a}^\dagger\hat{a}, \hat{\rho}\}\right) + g^2 n_\beta\left(\hat{a}^\dagger\hat{\rho}\hat{a} - \frac{1}{2}\{\hat{a}\hat{a}^\dagger, \hat{\rho}\}\right). \tag{12.14}$$

We used the identities $\mathrm{tr}(\hat{a}^\dagger\hat{a}\hat{\rho}_\beta) = n_\beta$ and $\mathrm{tr}(\hat{a}\hat{a}^\dagger\hat{\rho}_\beta) = 1 - n_\beta$, cf. Sect. 12.1. Note the appearance of a Lindblad structure (Sect. 8.6). The expression of $\mathrm{d}\hat{\rho}/\mathrm{d}t$ consists of the Hamiltonian part and of the above perturbative contribution of the collisions repeated at frequency $\nu$:

$$\frac{\mathrm{d}\hat{\rho}}{\mathrm{d}t} = -\frac{\mathrm{i}}{\hbar}[\hat{H}, \hat{\rho}] + \nu g^2(1-n_\beta)\left(\hat{a}\hat{\rho}\hat{a}^\dagger - \frac{1}{2}\{\hat{a}^\dagger\hat{a}, \hat{\rho}\}\right) + \nu g^2 n_\beta\left(\hat{a}^\dagger\hat{\rho}\hat{a} - \frac{1}{2}\{\hat{a}\hat{a}^\dagger, \hat{\rho}\}\right). \tag{12.15}$$

This perturbative result becomes exact in the limit $g \to 0, \nu \to \infty$ at $\nu g^2 =$ const. Let us cast it into the standard form, introduce the constant $\Gamma = \nu g^2(1-n_\beta)$ and consider that $n_\beta = \mathrm{e}^{-\beta\epsilon}(1-n_\beta)$. In such a way we obtain the form (8.25) of the thermalization master equation:

$$\frac{\mathrm{d}\hat{\rho}}{\mathrm{d}t} \equiv \mathcal{L}\hat{\rho} = -\frac{\mathrm{i}}{\hbar}[\hat{H}, \hat{\rho}] + \Gamma\left(\hat{a}\hat{\rho}\hat{a}^\dagger - \frac{1}{2}\{\hat{a}^\dagger\hat{a}, \hat{\rho}\}\right) + \mathrm{e}^{-\beta\epsilon}\Gamma\left(\hat{a}^\dagger\hat{\rho}\hat{a} - \frac{1}{2}\{\hat{a}\hat{a}^\dagger, \hat{\rho}\}\right). \tag{12.16}$$

The second term corresponds to spontaneous decay $|1\rangle \longrightarrow |0\rangle$ at rate $\Gamma$. The third term corresponds to thermal excitation $|0\rangle \longrightarrow |1\rangle$ at rate suppressed by the usual Boltzmann factor. The competition between the decay and excitation (emission and absorption) leads to the state $\hat{\rho}_\beta$ whose stationarity can be confirmed by substitution into the master equation above.

## 12.5 Q-Refrigerator

We are going to present a refrigerator[2] which operates on two qubits only, in contact with two respective thermal reservoirs—a cold one at temperature $T_c$ and a hot one at temperature $T_h > T_c$. The refrigerator will develop a colder part than the cold reservoir, able to cool further qubits to a certain $T_0$ below $T_c$. We know from

[2] After [2] by Linden et al.

thermodynamics that it is indeed possible to build refrigerators with no moving parts and with no external resources of energy other than the heat flow from a hot reservoir to a cold one. The peculiarity of our q-refrigerator is that it can operate on a single two-qubit system.

Suppose the first qubit of excitation energy $\epsilon_c$ is in contact with the cold reservoir, the second qubit of a *smaller* excitation energy $\epsilon_h < \epsilon_c$ is in contact with the hot reservoir. Both qubits will reach their thermal equilibrium states (12.1) of inverse temperatures $\beta_c = (1/k_B T_c)$ and $\beta_h = (1/k_B T_h)$, respectively:

$$\begin{aligned} \hat{\rho}_c &= \frac{|0;c\rangle\langle 0;c| + \exp(-\beta_c\epsilon_c)|1;c\rangle\langle 1;c|}{1+\exp(-\beta_c\epsilon_c)}, \\ \hat{\rho}_h &= \frac{|0;h\rangle\langle 0;h| + \exp(-\beta_h\epsilon_h)|1;h\rangle\langle 1;h|}{1+\exp(-\beta_h\epsilon_h)}. \end{aligned} \tag{12.17}$$

The refrigerator itself consists of these two qubits forming a four-state q-system. The energy spectrum is $0 < \epsilon_h < \epsilon_c < \epsilon_h + \epsilon_c$. The corresponding four eigenstates are

$$\begin{aligned} |0\rangle &\equiv |0;c\rangle \otimes |0;h\rangle, \\ |\epsilon_h\rangle &\equiv |0;c\rangle \otimes |1;h\rangle, \\ |\epsilon_c\rangle &\equiv |1;c\rangle \otimes |0;h\rangle, \\ |\epsilon_h+\epsilon_c\rangle &\equiv |1;c\rangle \otimes |1;h\rangle. \end{aligned} \tag{12.18}$$

The composite state of the refrigerator is $\hat{\rho}_h \otimes \hat{\rho}_c$ which yields

$$\frac{|0\rangle\langle 0| + \exp(-\beta_h\epsilon_h)|\epsilon_h\rangle\langle\epsilon_h| + \exp(-\beta_c\epsilon_c)|\epsilon_c\rangle\langle\epsilon_c| + \exp(-\beta_h\epsilon_h - \beta_c\epsilon_c)|\epsilon_h+\epsilon_c\rangle\langle\epsilon_h+\epsilon_c|}{1+\exp(-\beta_h\epsilon_h)+\exp(-\beta_c\epsilon_c)+\exp(-\beta_h\epsilon_h-\beta_c\epsilon_c)}. \tag{12.19}$$

We concentrate our attention on the two-dimensional subspace $\{|\epsilon_h\rangle, |\epsilon_c\rangle\}$ whose excitation energy is $\epsilon_0 = \epsilon_c - \epsilon_h$. We can see that the population of the "excited" state $|\epsilon_c\rangle$ is suppressed by a factor $\exp(-\beta_c\epsilon_c + \beta_h\epsilon_h)$ w.r.t. the "ground" state $|\epsilon_h\rangle$. This suppression would correspond to thermal equilibrium at a certain temperature $T_0$ which we can easily determine if we identify the above factor of suppression by the Boltzmann factor $\exp(-\beta_0\epsilon_0)$ with notation $\beta_0 = (1/k_B T_0)$. This way we obtain the constraint $\beta_c\epsilon_c - \beta_h\epsilon_h = \beta_0\epsilon_0$. It is trivial to solve it for $T_0$:

$$T_0 = \frac{1-(\epsilon_h/\epsilon_c)}{1-(T_c/T_h)(\epsilon_h/\epsilon_c)} T_c, \tag{12.20}$$

which is smaller than $T_c$ since we supposed that $T_c < T_h$ and $\epsilon_c > \epsilon_h$.

This means that the subspace $\{|\epsilon_h\rangle, |\epsilon_c\rangle\}$ is colder than the cold reservoir, the refrigerator can cool a thermometer, i.e., an additional qubit to temperature $T_0$ below $T_c$. The qubit to be cooled must have the same excitation energy $\epsilon_0$ which the transition $|\epsilon_h\rangle \to |\epsilon_c\rangle$ has, its coupling to the subspace $\{|\epsilon_h\rangle, |\epsilon_c\rangle\}$ must preserve the energy, and must be weak w.r.t. the thermalization rates of both the hot and cold reservoirs, respectively.

## 12.6 Thermal Qubit with External Work

In addition to heat exchange, thermodynamics treats energy exchange in the form of work as well. This concept applies to qubit-thermodynamics whenever we can externally control the Hamiltonian of the qubit. This happens most typically for the Hamiltonian $\hat{H} = -\frac{1}{2}\hbar\boldsymbol{\omega}\hat{\boldsymbol{\sigma}}$ (5.24) of an electronic or nuclear spin in external magnetic field $\boldsymbol{\omega}$. We restrict our discussion for the special case where the excitation energy $\epsilon$ of the qubit can be deliberately controlled by the external vertical magnetic field $\omega_z = \omega \geq 0$:

$$\epsilon = \hbar\omega. \tag{12.21}$$

Our previous results on the thermal qubit remain valid with one exception. We do not shift the ground state energy to zero, rather we retain the eigenvalues $\pm\frac{1}{2}\epsilon = \pm\frac{1}{2}\hbar\omega$ for the excited/ground states because we are interested in their variation with the magnetic field.

Suppose we prepare an equilibrium state (12.1) at a certain (inverse) temperature $\beta$ and magnetic field $\omega$:

$$\hat{\rho}_{\beta,\omega} = \frac{1}{2\cosh(\beta\hbar\omega/2)} e^{\beta\hbar\omega\hat{\sigma}_z/2}. \tag{12.22}$$

The average energy is the expression (12.3) minus $\frac{1}{2}\epsilon = \frac{1}{2}\hbar\omega$:

$$E = \mathrm{tr}(\hat{H}\hat{\rho}_{\beta,\omega}) = -\frac{\hbar\omega}{2}\sinh(\beta\hbar\omega/2), \quad -\frac{1}{2}\hbar\omega \leq 0. \tag{12.23}$$

What happens if we change the magnetic field with time and the Hamiltonian becomes time dependent: $\hat{H}(t) = -\frac{1}{2}\hbar\omega(t)\hat{\sigma}_z$? The change of the average energy has two contributions: the work $W$ exerted on the qubit by the variation of the magnetic field, and the heat $Q$ flowing from the reservoir into the qubit. Accordingly:

$$\frac{\mathrm{d}E}{\mathrm{d}t} = \mathrm{tr}\left(\frac{\mathrm{d}\hat{H}}{\mathrm{d}t}\hat{\rho}\right) + \mathrm{tr}\left(\hat{H}\frac{\mathrm{d}\hat{\rho}}{\mathrm{d}t}\right) \equiv \frac{\mathrm{d}W}{\mathrm{d}t} + \frac{\mathrm{d}Q}{\mathrm{d}t}, \tag{12.24}$$

where $\mathrm{d}W/\mathrm{d}t$ is the power and $\mathrm{d}Q/\mathrm{d}t$ is the rate of heat flow.[3] They depend on the dynamics of the qubit, which has two special cases of interest: the isothermal and the isentropic (or adiabatic[4]) processes.

In the isothermal process we leave the qubit in contact with the thermal reservoir and keep $\mathrm{d}\omega/\mathrm{d}t$ much smaller than the thermalization rate $\Gamma$ (cf. Sect. 12.4). Hence the qubit remains always thermalized at constant temperature for different

---

[3] Cf. [3] by Alicki, [4] by Spohn and Lebowitz.

[4] Thermodynamic adiabaticity, meaning "no heat transfer", differs from dynamical adiabaticity, a synonym of being quasi-static, which, e.g., in Chap. 3 stands for "no excitation".

magnetic fields: $\hat{\rho}(t) = \hat{\rho}_{\beta,\omega(t)}$ will be a good approximation. The power and the heat flow rate take the following forms, respectively:

$$\begin{aligned} \frac{\mathrm{d}W}{\mathrm{d}t} &= \mathrm{tr}\left(\frac{\mathrm{d}\hat{H}}{\mathrm{d}t}\hat{\rho}\right) = -\frac{\hbar}{2}\frac{\mathrm{d}\omega}{\mathrm{d}t}\sinh(\beta\hbar\omega/2) \\ \frac{\mathrm{d}Q}{\mathrm{d}t} &= \mathrm{tr}\left(\hat{H}\frac{\mathrm{d}\hat{\rho}}{\mathrm{d}t}\right) = -\frac{\beta\hbar\omega}{4}\hbar\frac{\mathrm{d}\omega}{\mathrm{d}t}\cosh(\beta\hbar\omega/2). \end{aligned} \tag{12.25}$$

In the isentropic process we isolate the qubit from the thermal reservoir. Without heat transfer, the slow variation of $\omega$ leaves the process reversible. In our special case $\hat{\rho}(t)$ remains the initial equilibrium state $\hat{\rho}_{\beta,\omega}$ because $[\hat{H}(t), \hat{\rho}_{\beta,\omega}] = 0$ for all $t$. The constancy of $\hat{\rho}$ makes the above defined heat flow vanish: $\mathrm{d}Q/\mathrm{d}t = 0$. That should indeed be so in isentropic processes. The only form of energy exchange is work:

$$\frac{\mathrm{d}W}{\mathrm{d}t} = \mathrm{tr}\left(\frac{\mathrm{d}\hat{H}}{\mathrm{d}t}\hat{\rho}(0)\right) = -\frac{\hbar}{2}\frac{\mathrm{d}\omega}{\mathrm{d}t}\sinh(\beta\hbar\omega(0)/2). \tag{12.26}$$

In both the isothermal and the isentropic cases the power $\mathrm{d}W/\mathrm{d}t$ is negative if $\mathrm{d}\omega/\mathrm{d}t$ is positive, i.e., an increasing magnetic field leads to work extraction from the qubit. (In the opposite case $\mathrm{d}\omega/\mathrm{d}t < 0$, the magnetic field performs work on the qubit.)

Due to the extreme simple structure of the qubit, the thermodynamics of its magnetization is degenerate. In common macroscopic magnetism, the equilibrium energy $E$ and magnetization $M$ are independent extensive variables. For the qubit, the average magnetization is defined as $M = \frac{1}{2}\mathrm{tr}(\hat{\sigma}_z\hat{\rho})$ which leads to the following constraint between $E$ and $M$:

$$E = -\hbar\omega M. \tag{12.27}$$

Therefore it makes sense to construct the thermodynamic entropy $S$ in function of $E$ or $M$ but not in both. Apart from this, the thermodynamics of the qubit is perfectly meaningful.

## 12.7 Q-Carnot Cycle

Beside its historic role in heat-engine theory, the Carnot cycle is of extreme theoretical value for thermodynamic reversibility. Heat $Q > 0$ from a hot reservoir of temperature $T_h$ can spontaneously go to a cold one of temperature $T_c < T_h$ but not vice versa. In thermodynamics, the transfer of heat $Q$ from $T_h$ decreases the entropy by $Q/T_c$ and the absorption of heat at $T_c$ increases the entropy by $Q/T_c > Q/T_h$ hence the net entropy change is positive whereas the opposite transfer would decrease the net entropy. Thermodynamics formulates the impossibility of the opposite transfer by the second law: entropy never decreases in a closed system.

Carnot cycle is a smart open system that is able to transfer heat reversibly (i.e., at constant entropy) between $T_h$ and $T_c$ in both directions. For example, it extracts heat $Q_h > 0$ from the hot reservoir and delivers heat $|Q_c| = \kappa Q_h$ to the cold one. The difference $L = Q_h - |Q_c|$ is turned into work performed by the machine on the environment. Here $\eta = L/Q_h = 1 - \kappa$ is the Carnot efficiency. If we re-supply this work into the machine, it will re-extract the heat from the cold reservoir, turn work into heat, and return both of them to the hot reservoir.

The Carnot machine consists of a thermal equilibrium system, called the working fluid, whose parameters are externally controlled. In our case, the working fluid is a single qubit and we control whether it is in contact with the hot or the cold reservoir or is just isolated, and we control the external magnetic field $\omega$ as well.[5] To pour heat from the hot reservoir into the cold one reversibly, the control means the following four processes repeated cyclically:

1. Isothermal "compression": the qubit is in contact with the hot reservoir and we slowly decrease the magnetic field from $\omega_1$ to $\omega_2 < \omega_1$. The qubit state varies from $\hat{\rho}_{\beta_h,\omega_1}$ to $\hat{\rho}_{\beta_h,\omega_2}$.
2. Adiabatic "compression": the qubit is isolated, the magnetic field is further decreased to $\kappa\omega_2$. The qubit state $\hat{\rho}_{\beta_h,\omega_2}$ is constant, but its temperature becomes $T_c$, hence we have to write it as $\hat{\rho}_{\beta_c,\kappa\omega_2}$.
3. Isothermal "expansion": the qubit is in contact with the cold reservoir and we slowly increase the magnetic field from $\kappa\omega_2$ to $\kappa\omega_1$. The qubit state varies from $\hat{\rho}_{\beta_c,\kappa\omega_2}$ to $\hat{\rho}_{\beta_c,\kappa\omega_1}$.
4. Adiabatic "expansion": the qubit is isolated, the magnetic field is further increased back to $\omega_1$. The qubit state $\hat{\rho}_{\beta_c,\kappa\omega_1}$ is constant, but its temperature becomes $T_h$, hence we have to write it as $\hat{\rho}_{\beta_h,\omega_1}$ which restores the initial state of the cycle.

Following Sect. 12.6, we calculate the integral heat and work consumptions, respectively, in each step in turn:

$$1.\ W_h = \frac{1}{\beta_h}\cosh\left(\frac{1}{2}\beta_h\hbar\omega\right)\Big|_{\omega_2}^{\omega_1} > 0, \qquad Q_h = \frac{\hbar}{2}\omega\sinh\left(\frac{1}{2}\beta_h\hbar\omega\right)\Big|_{\omega_2}^{\omega_1} - W_h > 0$$

$$2.\ W_{hc} = \eta\kappa\frac{\hbar}{2}\omega_2\cosh\left(\frac{1}{2}\beta_h\hbar\omega\right)\Big|_{\omega_1}^{\omega_2} < 0, \quad Q_{hc} = 0$$

$$3.\ W_c = \frac{1}{\beta_c}\cosh\left(\frac{1}{2}\beta_c\hbar\omega\right)\Big|_{\kappa\omega_1}^{\kappa\omega_2} < 0, \qquad Q_c = \frac{\hbar}{2}\omega\sinh\left(\frac{1}{2}\beta_c\hbar\omega\right)\Big|_{\kappa\omega_1}^{\kappa\omega_2} - W_c < 0$$

$$4.\ W_{ch} = \eta\kappa\frac{\hbar}{2}\omega_1\cosh\left(\frac{1}{2}\beta_h\hbar\omega\right)\Big|_{\omega_2}^{\omega_1} > 0, \quad Q_{ch} = 0.$$

In each cycle, the qubit absorbs the heat $Q_h > 0$ from the hot reservoir, delivers the work $L = -W_h - W_{hc} - W_c - W_c > 0$ to the magnetic field, and loses the heat $|Q_c| = Q_h - L$ to the cold reservoir. Running the cycle the other way around, these

[5] cf. [5] by Geva and Kosloff.

quantities change sign and heat will be pumped from the cold to the hot reservoir at the expense of work performed by the magnetic field on the qubit. Recalling that $\kappa = \beta_h/\beta_c = 1 - \eta$, the above expressions confirm the Carnot efficiency:

$$\frac{L}{Q_h} = \eta. \tag{12.28}$$

The Carnot cycle is absolutely significant for macroscopic thermodynamic reversibility: heat can be moved back-and-forth at constant thermodynamic entropy, all we need is external mechanical work. The Carnot cycle's significance for microscopic quantum reversibility is far from being fully understood.

## 12.8 Problems, Exercises

12.1 *Terabyte equivalent to Joule/degree.* Consider 1 g water at room temperature and suppose we heat it up by 1°. Its informatic (von Neumann) entropy increases by an incredible large number of bits. Let us calculate the order of water's mass, necessarily a microscopic portion of 1 g, whose informatic entropy would increase just by one terabyte—a storage available for everyone's home computer now.

12.2 *Landauer's principle.*[6] The erasure of one bit information is accompanied by the release of at least $k_B T \ln 2$ of heat. Let us argue for the principle. Method: consider a q-storage in thermal environment.

12.3 *Universal thermalizer?*. One would ask whether the Lindblad structure

$$\Gamma\left(\hat{a}\hat{\rho}\hat{a}^\dagger - \frac{1}{2}\{\hat{a}^\dagger\hat{a},\hat{\rho}\}\right) + e^{-\beta\epsilon}\Gamma\left(\hat{a}^\dagger\hat{\rho}\hat{a} - \frac{1}{2}\{\hat{a}\hat{a}^\dagger,\hat{\rho}\}\right)$$

on the r.h.s of Eq. (12.16) corresponds to a universal thermalizer, whether it models the influence of the thermal reservoir for any qubit? Suppose a different Hamiltonian $\hat{H} = \epsilon'\hat{n}$ instead of $\hat{H} = \epsilon\hat{n}$ and find the stationary state.

12.4 *Mechanical work on a qubit.* Let us show, for pedagogical purposes, that the work performed by the variation of the external magnetic field $\omega$ could be equivalently performed by an external mechanical force.

12.5 *Adiabatic demagnetization.* We apply a strong magnetic field $\omega$ to our qubit at environmental temperature $T$ and wait until the qubit becomes thermalized. Then we suddenly decrease the magnetic field to a much lower value $\omega' \ll \omega$. What happens to the temperature of the qubit?

---

[6] Cf. [6] by Landauer.

# References

1. Scarani, V., Ziman, M., Štelmanovič, P., Gisin, N., Bužek, V.: Phys. Rev. Lett. **88**, 097905 (2002)
2. Linden, N., Popescu, S., Skrzypczyk, P.: Phys. Rev. Lett. **105**, 130401 (2010)
3. Alicki, R.: J. Phys. **12**, 103 (1979)
4. Spohn, H., Lebowitz, J.L.: Adv. Chem. Phys. **38**, 109 (1979)
5. Geva, E., Kosloff, R.: J. Chem. Phys. **96**, 3054 (1992)
6. Landauer, R.: IBM J. Res. Dev. **5**, 183 (1961)

# Appendix

How reliable is the principle (cf. Sect. 12.3) that identifies informatic and thermodynamic entropies? We present an example[1] where this principle, applied to both thermodynamic and q-informatic analysis of a simple model of irreversibility, will lead to an independent new—and exact—feature of q-entropy.

## A.1 Introduction

Consider a thermal reservoir at (inverse) temperature $\beta$. Thermodynamics determines its thermodynamic entropy $S^R_{\mathrm{th}}$. If $\hat{\rho}_R$ is the Gibbs canonical density matrix of the reservoir, we can calculate the informatic (von Neumann) entropy as well. We can, in principle, prove that the thermodynamic entropy $S^R_{\mathrm{th}}$ of our equilibrium reservoir is equal to the informatic entropy:

$$S^R_{\mathrm{th}} = S(\hat{\rho}_R) \equiv -\mathrm{tr}(\hat{\rho}_R \log \hat{\rho}_R), \tag{A.1}$$

provided we take the so-called thermodynamic, i.e., infinite volume limit. (Units in Appendix are chosen such that $k_B \ln 2 = 1$, to count both the thermodynamic and informatic entropies in bits.) What happens if we move the system out of its equilibrium? Suppose we switch on a certain external macroscopic field for a while, then we switch it off. This perturbation will always increase the energy of the reservoir, the field will always perform work $W > 0$ on it. From the thermodynamic viewpont, part of this work $W$ will be dissipated into heat in the reservoir. We are interested in those situations where the whole of $W$ gets dissipated. Then the increase of the thermodynamic entropy of the reservoir is of the standard form:

$$\Delta S^R_{\mathrm{th}} = \beta W > 0. \tag{A.2}$$

[1] Cf. [1] by Diósi, Feldmann and Kosloff.

According to the principle discussed in Sect. 12.3 we might expect that this irreversible thermodynamic entropy production equals the change of the informatic entropy of the reservoir. If $\hat{\rho}'_R$ stands for the state after the external perturbation, we might expect the following equation to hold:

$$\Delta S^R_{\text{th}} = S(\hat{\rho}'_R) - S(\hat{\rho}_R). \tag{A.3}$$

But it cannot, since the external macroscopic field makes $\hat{\rho}_R$ evolve unitarily, reversibly. The perturbed state is of the form:

$$\hat{\rho}'_R = \hat{U}_R \hat{\rho}_R \hat{U}^\dagger_R, \tag{A.4}$$

with unitary $\hat{U}_R$, hence the r.h.s. of (A.3) is always zero. This is the notorious conflict between the reversibility of the reservoir's microscopic dynamics and the macroscopically observed irreversible dissipation. Resolutions of this contradiction can be based on a smart completion of the unitary evolution by a subsequent irreversible q-operation $\mathcal{M}$, then we might expect that the modified equation holds:

$$\Delta S^R_{\text{th}} = S(\mathcal{M}\hat{\rho}'_R) - S(\hat{\rho}_R). \tag{A.5}$$

Unfortunately, typical dissipative systems are so complex that we cannot fit our choice of $\mathcal{M}$ to satisfy this equation. Either the thermodynamic calculation of $\Delta S^R_{\text{th}}$ or the microscopic calculation of $S(\mathcal{M}\hat{\rho}'_R)$ or both are unavailable. Therefore we choose a specific strategy.

We consider the simplest ever thermal reservoir, which is the ideal qubit gas of Sect. 12.2, and we consider the simplest external perturbation, which is the single-qubit unitary map. This way we reduce technical difficulties to the minimum, the terms of Eq. A.5 become tractable technically. The equation becomes a powerful criterion to single out the irreversible q-operation $\mathcal{M}$.

## A.2 The Reservoir, Collisions

Consider a single qubit in constant magnetic field, with a Hamiltonian $\hat{H}$, and assume that it is initially in the Gibbs equilibrium state $\hat{\rho} = \hat{\rho}_\beta$ (12.1) at the inverse temperature $\beta$:

$$\hat{\rho} = \mathcal{N} e^{-\beta\hat{H}}. \tag{A.6}$$

Assume an arbitrary short pulse superposed on the constant magnetic field to cause a 'collision' to the qubit state:

$$\hat{\rho} \to \hat{U}\hat{\rho}\hat{U}^\dagger \equiv \hat{\sigma}, \tag{A.7}$$

where $\hat{U}$ is unitary since the dynamics of the qubit is Hamiltonian all along the pulse. Suppose our thermal reservoir is formed by the ideal gas of $n$ distinguishable qubits (molecules) in equilibrium:

$$\hat{\rho}_R = \hat{\rho} \otimes \hat{\rho} \otimes \ldots \hat{\rho} \equiv \hat{\rho}^{\otimes n}, \tag{A.8}$$

with the total Hamiltonian

$$\hat{H}_R = \hat{H} \otimes \hat{I}^{\otimes(n-1)} + \hat{I} \otimes \hat{H} \otimes \hat{I}^{\otimes(n-2)} + \hat{I}^{\otimes(n-1)} \otimes \hat{H}, \tag{A.9}$$

and its thermodynamic limit is $n \to \infty$. Let us assume that the qubits of the reservoir will reversibly collide with the pulses of the field according to the Eq. A.7. Without restricting the generality, we can assume that the 1st qubit collides first, the 2nd collides second, etc.:

$$\hat{\rho}^{\otimes n} \to \hat{\sigma} \otimes \hat{\rho}^{\otimes(n-1)} \to \hat{\sigma} \otimes \hat{\sigma} \otimes \hat{\rho}^{\otimes(n-2)} \to \ldots \tag{A.10}$$

Since $S(\hat{\sigma}) = S(\hat{\rho})$, the informatic entropy of the reservoir cannot change at all in the above reversible collisions:

$$S(\hat{\rho}^{\otimes n}) = S(\hat{\sigma} \otimes \hat{\rho}^{\otimes(n-1)}) = S(\hat{\sigma} \otimes \hat{\sigma} \otimes \hat{\rho}^{\otimes(n-2)}) = \ldots = nS(\hat{\rho}) = nS(\hat{\sigma}). \tag{A.11}$$

For simplicity, we consider the first collision:

$$\hat{\rho}'_R = \hat{\sigma} \otimes \hat{\rho}^{\otimes(n-1)}, \tag{A.12}$$

and implement our idea for it. We define the average work $W$ of the field pulse performed on the qubit:

$$W \equiv \mathrm{tr}(\hat{H}\hat{\sigma}) - \mathrm{tr}(\hat{H}\hat{\rho}). \tag{A.13}$$

We express $\hat{H} = (\hat{I} \log \mathcal{N} - \log \hat{\rho})/\beta$ from (A.6), substitute it, and observe that $\mathrm{tr}(\hat{\rho} \log \hat{\rho}) = \mathrm{tr}(\hat{\sigma} \log \hat{\sigma})$:

$$W = \frac{\mathrm{tr}(\hat{\sigma} \log \hat{\sigma}) - \mathrm{tr}(\hat{\sigma} \log \hat{\rho})}{\beta} \equiv \frac{S(\hat{\sigma}\|\hat{\rho})}{\beta}, \tag{A.14}$$

We have recognized the appearance of the relative q-entropy (A.9) of the post- and pre-collision states. Since the qubit is part of a reservoir and the field interacts with many qubits in succession then, thermodynamically, we postulate that the above energy (the work by the pulses) is dissipated to the reservoir. Therefore, the average thermodynamic entropy production per collision is $\Delta S^R_{\mathrm{th}} = \beta W$, which using (A.14), reads:

$$\Delta S^R_{\mathrm{th}} = S(\hat{\sigma}\|\hat{\rho}). \tag{A.15}$$

We see that $\Delta S^R_{\mathrm{th}}$ is always positive since $S(\hat{\sigma}\|\hat{\rho})$ is always positive if $\hat{\sigma} \neq \hat{\rho}$. The above identity is part of standard statistical physics.

Our target identity is different, it is just (A.5) expressing the identity of the thermodynamic and informatic entropy productions, respectively. Let us substitute the microscopic expression (A.15) of $\Delta S_{\rm th}^R$, the expression (A.8) of $\hat\rho_R$, and the expression (A.12) of $\hat\rho'_R$. We get the following equation between various von Neumann entropies:

$$S(\hat\sigma\|\hat\rho) = S(\mathcal{M}\hat\sigma\otimes\hat\rho^{\otimes(n-1)}) - S(\hat\sigma\otimes\hat\rho^{\otimes(n-1)}). \tag{A.16}$$

We expect this identity to hold in the thermodynamic limit $n\to\infty$. We have yet to specify the irreversible q-operation $\mathcal{M}$.

## A.3 The Graceful Irreversible Operation

We must postulate an irreversible q-operation $\mathcal{M}$ which satisfies the constraint (A.16) while it does but gracefully modify the exact post-collision state $\hat\sigma\otimes\hat\rho^{\otimes(n-1)}$. The map

$$\hat\sigma\otimes\hat\rho^{\otimes(n-1)} \to \mathcal{M}\left(\hat\sigma\otimes\hat\rho^{\otimes(n-1)}\right) \tag{A.17}$$

should obviously be symmetric for the permutation of the qubits. Single-qubit maps can increase the informatic entropy but they are not likely to produce the requested value $S(\hat\sigma\|\hat\rho)$. Hence $\mathcal{M}$ should correlate the qubits. On the other hand, $\mathcal{M}$ should only smear out information whose loss is heuristically justifiable in a reservoir like ours. It must not change the total energy or the total magnetization. Interestingly, there is a beautiful simple choice: let $\mathcal{M}$ be, in general, the total symmetrization over all permutations of the $n$ qubits. In our case it means:

$$\mathcal{M}\left(\hat\sigma\otimes\hat\rho^{\otimes(n-1)}\right) = \frac{\hat\sigma\otimes\hat\rho^{\otimes(n-1)} + \hat\rho\otimes\hat\sigma\otimes\hat\rho^{\otimes(n-2)} + \cdots + \hat\rho^{\otimes(n-1)}\otimes\hat\sigma}{n}. \tag{A.18}$$

It is clear that this post-collision operation is irreversible and increases the informatic entropy of the reservoir. It is still an open issue if it produces as much entropy $S(\hat\sigma\|\hat\rho)$ as expected thermodynamically.

## A.4 New Math Conjecture on Relative Q-Entropy

Let us summarize what we have done so far. For an ideal qubit gas we considered single qubit reversible perturbations by repeated external pulses. We postulated that the work by the pulses is totally dissipated in the macroscopic thermodynamic sense. This way we postulated the amount of thermodynamic entropy production per pulses. To produce microscopic irreversibility within the model, we imposed a graceful irreversible operation which we think would produce as much von

Neumann entropy as thermodynamics did. We found that this latter condition depends on the validity of the following mathematical conjecture:

$$\lim_{n=\infty}\left(S(\mathcal{M}(\hat{\sigma}\otimes\hat{\rho}^{\otimes(n-1)}))-S(\hat{\sigma}\otimes\hat{\rho}^{\otimes(n-1)})\right)=S(\hat{\sigma}\|\hat{\rho}), \tag{A.19}$$

where $\mathcal{M}$ is full symmetrization (A.18).

The derivation of the conjecture involved numerous heuristic steps, our choice of the irreversible operation $\mathcal{M}$ was purely intuitive, and the heart of the derivation was the principle that thermodynamic and informatic entropies must coincide even along irreversible processes. So, the conjecture could easily have proven wrong. But it is correct [2], we got a new independent theorem and interpretation for the relative q-entropy.

## References

1. Diósi, L., Feldmann, T., Kosloff, R.: Int. J. Quant. Inf. **4**, 99 (2006)
2. Csiszár, I., Hiai, F., Petz, D.: J. Math. Phys. **48**, 092102 (2007)

# Solutions

## Problems of Chap. 2

2.1 *Mixture of pure states.* It is easy to see that $\rho(\mathrm{x}) = \rho(0)\delta_{\mathrm{x}0} + \rho(1)\delta_{\mathrm{x}1}$ which suggests that the distribution of the mixing weights must coincide with the distribution $\rho$. Accordingly, in the case of continuous phase space, we can choose $w(\bar{x}) = \rho(\bar{x})$ to mix the pure states $\delta(x - \bar{x})$:

$$\rho(x) = \int w(\bar{x})\delta(x - \bar{x})\mathrm{d}\bar{x}.$$

2.2 *Probabilistic or non-probabilistic mixing?* Mixing $n$ zeros and $n$ ones is realized by their random permutation that amounts to the following $2n$-partite composite state:

$$\rho(\mathrm{x}_1, \ldots, \mathrm{x}_{2n}) = \frac{\delta_{\mathrm{x}_1 0} \ldots \delta_{\mathrm{x}_n 0}\delta_{\mathrm{x}_{n+1} 1} \ldots \delta_{\mathrm{x}_{2n} 1} + \text{permutations of } \mathrm{x}_1, \ldots, \mathrm{x}_{2n}}{\#\text{ of permutations}}.$$

In case of the probabilistic mixing, however, we make $2n$ random decisions as to choose a zero or a one and then we mix the $2n$ elements. The probabilistic mixing amounts to $\rho(\mathrm{x}_1) \ldots \rho(\mathrm{x}_{2n})$ which is obviously different from the above composite state.

2.3 *Classical separability.* If we choose $w(\bar{x}_A, \bar{x}_B) = \rho_{AB}(\bar{x}_A, \bar{x}_B)$ and replace summation over the weights by integration, it works:

$$\rho_{AB}(x_A, x_B) = \int w(\bar{x}_A, \bar{x}_B)\delta(x_A - \bar{x}_A)\delta(x_B - \bar{x}_B)\mathrm{d}\bar{x}_A\mathrm{d}\bar{x}_B.$$

Thus all $\rho_{AB}$ are mixtures of the uncorrelated pure states $\delta(x_A - \bar{x}_A)\delta(x_B - \bar{x}_B)$.

2.4 *Decorrelating a single state?* The map $\rho_{AB} \rightarrow \rho_A\rho_B$ is nonlinear:

$$\rho_{AB}(x_A, x_B) \longrightarrow \int \rho_{AB}(x_A, x'_B)\mathrm{d}x'_B \int \rho_{AB}(x'_A, x_B)\mathrm{d}x'_A.$$

Hence the map is not a real operation.

2.5 *Decorrelating an ensemble*. The collective state $\rho_{AB}^{\times 2n}$ is granted to start with. We interchange the first $n$ and the second $n$ subsystems A. Then we trace over the second $n$ composite systems AB. We get $(\rho_A\rho_B)^{\times n}$. This can be verified if, e.g., we calculate the expectation values of observables like $f(x_{A_1})g(x_{B_1})$, and $f_1(x_{A_1})f_2(x_{A_2})g_1(x_{B_1})g_2(x_{B_2})$, etc.

2.6 *Measurement and Bayes theorem*. The Bayes theorem states that the a priori distribution $\rho(x)$ will be updated if we learn the value $n$ of a variable which was a priori correlated with $x$. To calculate the updated conditional distribution $\rho(x|n)$ from the a priori $\rho(x)$, we need to know the a priori distribution $\rho(n)$ of $n$ as well as its conditional distribution $\rho(n|x)$:

$$\rho(x|n) = \frac{\rho(n|x)\rho(x)}{\rho(n)}.$$

The quantities $\rho(x|n)$ and $\rho(n)$ correspond to $\rho_n(x)$ and $p_n$, respectively, in the notations of the measurement scheme of Sect. 2.4. The above Bayesian prediction becomes completely identical with the prediction $\rho_n(x)$ of the measurement (2.20) if we identify the measured effects by the Bayesian conditional distributions of $n$:

$$\Pi_n(x) = \rho(n|x).$$

2.7 *Indirect measurement*. Imagine a detector of discrete state space $\{n\}$. Let the composite state of the system and the detector be $\rho(x, n) = \rho(x)\Pi_n(x)$ where $\rho(x)$ stands for the reduced state of the system. Observe that $\Pi_n(x)$ becomes the conditional state $\rho(n|x)$ of the detector for the system being in the pure state $x$. Now we perform a projective measurement on the detector quantity $n$. Formally, the partition $\{P_m\}$ of the detector state space must be defined as $P_m(n) = \delta_{mn}$ for all $m$. According to the rules of projective measurement, the state change will be

$$\rho(x,n) \to \rho_m(x,n) \equiv \frac{1}{p_m}\delta_{mn}\rho(x,n),$$

with probability $p_m = \int \rho(x,m)\mathrm{d}x$. For the reduced state $\rho(x) = \sum_m \rho(x,m)$, the above projective measurement induces the desired non-projective measurement of the effects $\{\Pi_n(x)\}$.

## Problems of Chap. 3

3.1 *Bohr quantization of the harmonic oscillator*. The sum of the kinetic and potential energies yields the total energy $E = \frac{1}{2}p^2 + \frac{1}{2}\omega^2 q^2$ which is constant during the motion. Therefore the phase space point $(q, p)$ moves on the ellipse of surface $2\pi E/\omega$. The surface plays a role in the Bohr–Sommerfeld q-condition because the contour-integral of $pdq$ for one period is equal to the surface of the enclosed ellipse. Hence the canonical action takes the form $E/\omega$ and we get

$$E/\omega = \hbar\left(n+\frac{1}{2}\right).$$

3.2 *The role of adiabatic invariants.* The canonical actions $I_k$ are adiabatic invariants of the classical motion. This means that they remain approximately constant against whatever large variations of the external parameters of the Hamilton-function provided the variations are slow with respect to the motion. So, the canonical action $I$ of the oscillator will be invariant against the variation of $\omega$ in the Hamiltonian $\frac{1}{2}p^2+\frac{1}{2}\omega^2(t)q^2$ provided $|\dot{\omega}| \ll \omega^2$. The q-condition remains satisfied with the same q-number $n$.

3.3 *Classical-like or q-like motion.* The Bohr–Sommerfeld q-condition restricts the continuum of classical motions to a discrete infinite sequence. For small q-numbers this restriction is relevant since the allowed phase space trajectories are well separated. For large q-numbers, typically, the allowed trajectories become quite dense in phase space and might fairly approximate any classical trajectory which does otherwise not satisfy the q-conditions.

## Problems of Chap. 4

4.1 *Decoherence-free projective measurement.* Let us construct the spectral expansion $\hat{A}=\sum_\lambda A_\lambda \hat{P}_\lambda$ and the post-measurement state $\hat{\rho}'=\sum_\lambda \hat{P}_\lambda\hat{\rho}\hat{P}_\lambda$. If $[\hat{A},\hat{\rho}]=0$ then $\hat{\rho}'=\hat{\rho}$ since $[\hat{A},\hat{\rho}]=0$ is equivalent with $[\hat{P}_\lambda,\hat{\rho}]=0$ for all $\lambda$. To prove the inverse statement, that $\hat{\rho}'=\hat{\rho}$ implies $[\hat{A},\hat{\rho}]=0$, we can simply write $[\hat{A},\hat{\rho}]=[\hat{A},\hat{\rho}']=0$.

4.2 *Mixing the eigenstates.* Let us consider the spectral expansion of the matrix $\hat{\rho}$:

$$\hat{\rho}=\sum_\lambda \rho_\lambda \hat{P}_\lambda.$$

If $\hat{\rho}$ is non-degenerate then the $\hat{P}_\lambda$'s correspond to the pure eigenstates of $\hat{\rho}$ and their mixture yields the state $\hat{\rho}$ if the corresponding eigenvalues make the mixing weights: $w_\lambda=\rho_\lambda$. In the general case, the spectral expansion implies the mixture $\hat{\rho}=\sum_\lambda w_\lambda\hat{\rho}_\lambda$ with $w_\lambda=d_\lambda\rho_\lambda$ and $\hat{\rho}_\lambda=\hat{P}_\lambda/d_\lambda$ where $d_\lambda$ is the dimension of $\hat{P}_\lambda$.

4.3 *Weak measurement of correlation.* We shall need both spectral decompositions $\hat{A}=\sum_\lambda A_\lambda\hat{P}_\lambda$ and $\hat{B}=\sum_\lambda B_\lambda\hat{Q}_\lambda$. After the unsharp measurement of $\hat{A}$ on the state $\hat{\rho}$, the post-measurement state becomes this:

$$\hat{\rho}_{\bar{A}}=\frac{1}{p_{\bar{A}}\sqrt{2\pi\sigma^2}}\exp\left[-\frac{(\bar{A}-\hat{A})^2}{4\sigma^2}\right]\hat{\rho}\exp\left[-\frac{(\bar{A}-\hat{A})^2}{4\sigma^2}\right],$$

where $p_{\bar{A}}$ is the probability of the measurement outcome $\bar{A}$. At the end of the day, we shall take the weak measurement limit $\sigma \to \infty$. Before it, we consider the projective measurement of $\hat{B}$ on the above post-measurement state, yielding the outcome $B_\lambda$ with probability $p_\lambda = \mathrm{tr}(\hat{Q}_\lambda \hat{\rho}_{\bar{A}})$. Then we can write

$$\langle \bar{A} B_\lambda \rangle = \int p_{\bar{A}} \bar{A} \sum_\lambda p_\lambda B_\lambda \mathrm{d}\bar{A} = \int p_{\bar{A}} \bar{A} \mathrm{tr}(\hat{B} \hat{\rho}_{\bar{A}}) \mathrm{d}\bar{A}.$$

We insert the expression of $\hat{\rho}_{\bar{A}}$, we get an integral on the r.h.s.:

$$\frac{1}{\sqrt{2\pi\sigma^2}} \int \bar{A} \exp\left[-\frac{(\bar{A} - \hat{A})^2}{4\sigma^2}\right] \hat{\rho} \exp\left[-\frac{(\bar{A} - \hat{A})^2}{4\sigma^2}\right] \mathrm{d}\bar{A},$$

which can be evaluated if we insert the spectral decomposition of $\hat{A}$; in the weak measurement limit it reduces to $\frac{1}{2}\{\hat{A}, \hat{\rho}\}$. Using this result, we obtain the desired equation $\langle \bar{A} B_\lambda \rangle = \frac{1}{2}\mathrm{tr}(\{\hat{A}, \hat{B}\} \hat{\rho})$.

4.4 *Separability of pure states.* If $|\psi_{AB}\rangle = |\psi_A\rangle \otimes |\psi_B\rangle$ then the composite density matrix $\hat{\rho}_{AB}$ is a single tensor product and it is trivially separable. The other way around, when the pure state satisfies the separability condition (4.48)

$$|\psi_{AB}\rangle\langle\psi_{AB}| = \sum_\lambda w_\lambda \hat{\rho}_{A\lambda} \otimes \hat{\rho}_{B\lambda},$$

then it follows that the matrices on both sides have rank 1. Accordingly, the r.h.s. must be equivalent to the tensor product of rank 1 (i.e.: pure state) density matrices:

$$|\psi_{AB}\rangle\langle\psi_{AB}| = |\psi_A\rangle\langle\psi_A| \otimes |\psi_B\rangle\langle\psi_B|,$$

which implies the form $|\psi_{AB}\rangle = |\psi_A\rangle \otimes |\psi_B\rangle$.

4.5 *Unitary cloning?* Let us suppose that we have duplicated two states $|\psi\rangle$ and $|\psi'\rangle$:

$$|\psi\rangle \otimes |\psi_0\rangle \longrightarrow |\psi\rangle \otimes |\psi\rangle; \quad |\psi'\rangle \otimes |\psi_0\rangle \longrightarrow |\psi'\rangle \otimes |\psi'\rangle.$$

The inner product of the two initial composite states is $\langle\psi|\psi'\rangle$ while the inner product of the two final composite states is $\langle\psi|\psi'\rangle^2$. Therefore the above process of state duplication cannot be unitary.

## Problems of Chap. 5

5.1 *Pure state fidelity from density matrices.* Observe that $\langle \boldsymbol{m}|\boldsymbol{n}\rangle^2$ equals the trace of the product $|\boldsymbol{n}\rangle\langle\boldsymbol{n}|$ times $|\boldsymbol{m}\rangle\langle\boldsymbol{m}|$. Let us invoke the Pauli-representation of these two density matrices and evaluate the trace of their product:

$$|\langle \boldsymbol{m}|\boldsymbol{n}\rangle|^2 = \mathrm{tr}\left(\frac{\hat{I}+\boldsymbol{n}\hat{\boldsymbol{\sigma}}}{2}\frac{\hat{I}+\boldsymbol{m}\hat{\boldsymbol{\sigma}}}{2}\right) = \frac{1+\boldsymbol{nm}}{2},$$

which yields $\cos^2(\vartheta/2)$.

5.2 *Unitary rotation for* $|\uparrow\rangle \longrightarrow |\downarrow\rangle$. Since $|\uparrow\rangle$ corresponds to the north pole and $|\downarrow\rangle$ corresponds to the south pole on the Bloch-sphere, we need a $\pi$-rotation around, e.g., the $x$-axis. The rotation vector is $\boldsymbol{\alpha} = (\pi, 0, 0)$ and the corresponding unitary transformation becomes

$$\hat{U}(\boldsymbol{\alpha}) \equiv \exp\left(-\frac{i}{2}\boldsymbol{\alpha}\hat{\boldsymbol{\sigma}}\right) = -\mathrm{i}\hat{\sigma}_x.$$

We can check the result directly:

$$-\mathrm{i}\hat{\sigma}_x|\uparrow\rangle = -\mathrm{i}\begin{bmatrix} 0 & 1 \\ 1 & 0 \end{bmatrix}\begin{bmatrix} 1 \\ 0 \end{bmatrix} = -\mathrm{i}\begin{bmatrix} 0 \\ 1 \end{bmatrix} = -\mathrm{i}|\downarrow\rangle.$$

5.3 *Density matrix eigenvalues and states in terms of polarization.* Consider the density matrix $\frac{1}{2}(\hat{I} + \boldsymbol{s}\hat{\boldsymbol{\sigma}})$ and find the spectral expansion of $\boldsymbol{s}\hat{\boldsymbol{\sigma}}$. We learned that if $\boldsymbol{s}$ is a unit vector then $\boldsymbol{s}\hat{\boldsymbol{\sigma}}|\uparrow s\rangle = |\uparrow s\rangle$ and $\boldsymbol{s}\hat{\boldsymbol{\sigma}}|\downarrow s\rangle = -|\downarrow s\rangle$. If $s \le 1$, the two eigenstates remain the same and we keep the simple notations $|\uparrow s\rangle, |\downarrow s\rangle$ to denote qubits polarized along or, respectively, opposite to the direction $\boldsymbol{s}$. The eigenvalues will change trivially and we have $\boldsymbol{s}\hat{\boldsymbol{\sigma}}|\uparrow s\rangle = s|\uparrow s\rangle$ and $\boldsymbol{s}\hat{\boldsymbol{\sigma}}|\downarrow s\rangle = -s|\downarrow s\rangle$. Then we can summarize the eigenvalues and eigenstates of the density matrix in the following way:

$$\frac{\hat{I}+\boldsymbol{s}\hat{\boldsymbol{\sigma}}}{2}|\uparrow s\rangle = \frac{1+s}{2}|\uparrow s\rangle, \quad \frac{\hat{I}+\boldsymbol{s}\hat{\boldsymbol{\sigma}}}{2}|\downarrow s\rangle = \frac{1-s}{2}|\downarrow s\rangle.$$

5.4 *Magnetic rotation for* $|\uparrow\rangle \longrightarrow |\downarrow\rangle$. We must implement a $\pi$-rotation of the polarization vector and we can choose the rotation vector $(\pi, 0, 0)$ which means $\pi$-rotation around the $x$-axis. In magnetic field $\boldsymbol{\omega}$, the polarization vector $\boldsymbol{s}$ satisfies the classical equation of motion $\dot{\boldsymbol{s}} = -\boldsymbol{\omega} \times \boldsymbol{s}$ meaning that $\boldsymbol{s}$ will rotate around the direction $\boldsymbol{\omega}$ of the field at angular velocity $\omega$. Accordingly, we can choose the field to point along the $x$-axis: $\boldsymbol{\omega} = (\omega, 0, 0)$. The rotation angle $\pi$ is achieved if we switch on the field for a period $t = \pi/\omega$.

5.5 *Interrelated qubit physical quantities.*

$$\hat{P}_n + \hat{P}_{-n} = \hat{I}$$
$$2\hat{P}_n - \hat{\sigma}_n = 2\hat{P}_{-n} + \hat{\sigma}_n = \hat{I}$$

5.6 *Mixing non-orthogonal polarizations.* Since the qubit density matrix is a linear function of the polarization vector, mixing the density matrices means averaging their polarization vectors with the mixing weights. Therefore our mixture has the following polarization vector:

$$s = \frac{1}{3} \times (0,0,1) + \frac{2}{3} \times (1,0,0) = (1/3, 0, 2/3).$$

## Problems of Chap. 6

6.1 *Universality of Hadamard and phase operations.* The unitary rotations $\hat{U}$ of the qubit space are, apart from an irrelevant phase, equivalent to the spatial rotations of the corresponding Bloch sphere. This time the three Euler angles $\psi$, $\theta$, $\phi$ are the natural parameters. We can write the unitary rotations, corresponding to the spatial ones, into this form:

$$\hat{U}(\psi,\theta,\phi) = \exp\left(-\frac{i}{2}\psi\hat{\sigma}_z\right)\exp\left(-\frac{i}{2}\theta\hat{\sigma}_x\right)\exp\left(-\frac{i}{2}\phi\hat{\sigma}_z\right).$$

The middle factor, too, becomes rotation around the $z$-axis if we sandwich it between two Hadamard operations because $\hat{H}\hat{\sigma}_x\hat{H} = \hat{\sigma}_z$. We can thus express the r.h.s. in the desired form

$$\hat{U}(\psi,\theta,\phi) = \hat{T}(\psi)\hat{H}\hat{T}(\theta)\hat{H}\hat{T}(\phi).$$

6.2 *Statistical error of qubit determination.* Out of $N$, we allocate $N_x$, $N_y$, $N_z$ qubits to estimate $s_x$, $s_y$, $s_z$, respectively. We learned that the estimated value of $s_x$ takes this form:

$$\frac{N_{\uparrow x} - N_{\downarrow x}}{N_{\uparrow x} + N_{\downarrow x}} = \frac{2N_{\uparrow x}}{N_x} - 1,$$

because on a large statistics $N_x = N_{\uparrow x} + N_{\downarrow x}$ the ratio $N_{\uparrow x}/N_x$ converges to the q-theoretical prediction $p_{\uparrow x} = \langle \uparrow x|\hat{\rho}|\uparrow x\rangle \equiv \frac{1}{2}(1+s_x)$. The statistical error of the estimation takes the form $2\Delta N_{\uparrow x}/N_x$ and we are going to determine the mean fluctuation $\Delta N_{\uparrow x}$. The statistical distribution of the count $N_{\uparrow x}$ is binomial:

$$p(N_{\uparrow x}) = \binom{N_x}{N_{\uparrow x}} p_{\uparrow x}^{N_{\uparrow x}} p_{\downarrow x}^{N_{\downarrow x}},$$

hence the mean squared fluctuation of the count $N_{\uparrow x}$ takes the form $(\Delta N_{\uparrow x})^2 = N_x p_{\uparrow x} p_{\downarrow x} = N_x(1-s_x^2)/4$. This yields the ultimate form of the estimation error:

$$\Delta s_x = \sqrt{\frac{1-s_x^2}{N_x}},$$

and we could get similar results for $\Delta s_y$ and $\Delta s_z$.

6.3 *Fidelity of qubit determination.* If the state $|\boldsymbol{n}\rangle$ sent by Alice and the polarization $\hat{\sigma}_m$ chosen by Bob were fixed then the structure of the expected fidelity of Bob's guess would be this:

$$|\langle\boldsymbol{n}|\boldsymbol{m}\rangle|^2 p_{\uparrow m} + |\langle\boldsymbol{n}|-\boldsymbol{m}\rangle|^2 p_{\downarrow m}.$$

Here we have understood that Bob's optimum guess must always be the post-measurement state $|\pm\boldsymbol{m}\rangle$ based on the measurement outcome $\hat{\sigma}_m = \pm 1$, respectively. Now we recall that $|\langle\boldsymbol{n}|\boldsymbol{m}\rangle|^2 = p_{\uparrow m} = \cos^2(\vartheta/2)$ where $\cos\vartheta = \boldsymbol{nm}$, and $|\langle\boldsymbol{n}|-\boldsymbol{m}\rangle|^2 = p_{\downarrow m} = \sin^2(\vartheta/2)$. Hence the above fidelity takes the simple form $\cos^4(\vartheta/2) + \sin^4(\vartheta/2)$ which we rewrite into the equivalent form $\frac{1}{2} + \frac{1}{2}\cos^2(\vartheta)$. The average of $\cos^2(\vartheta) = (\boldsymbol{nm})^2$ over the random independent $\boldsymbol{n}$ and $\boldsymbol{m}$ yields 1/3 therefore the expected fidelity of Bob's guess becomes 2/3.

6.4 *Post-measurement depolarization.* Let $\hat{\sigma}_n$ denote the polarization chosen by Bob. The non-selective measurement induces the change $\hat{\rho} \to \hat{P}_n\hat{\rho}\hat{P}_n + \hat{P}_{-n}\hat{\rho}\hat{P}_{-n}$ of the state. Inserting the Pauli-representation of $\hat{\rho}$ and the projectors $\hat{P}_{\pm n}$ yields

$$\frac{\hat{I}+\boldsymbol{s}\hat{\boldsymbol{\sigma}}}{2} \to \frac{\hat{I}+\hat{\sigma}_n}{2}\frac{\hat{I}+\boldsymbol{s}\hat{\boldsymbol{\sigma}}}{2}\frac{\hat{I}+\hat{\sigma}_n}{2} + \frac{\hat{I}-\hat{\sigma}_n}{2}\frac{\hat{I}+\boldsymbol{s}\hat{\boldsymbol{\sigma}}}{2}\frac{\hat{I}-\hat{\sigma}_n}{2}$$
$$= \frac{\hat{I}+\boldsymbol{s}\hat{\boldsymbol{\sigma}}}{4} + \frac{\hat{I}+\boldsymbol{s}\hat{\sigma}_n\hat{\boldsymbol{\sigma}}\hat{\sigma}_n}{4}.$$

Since Bob's choice is random regarding $\boldsymbol{n}$ we shall average $\boldsymbol{n}$ over the solid angle. Averaging the structure $\hat{\sigma}_n\hat{\boldsymbol{\sigma}}\hat{\sigma}_n$ yields $-\hat{\boldsymbol{\sigma}}/3$ hence the average influence of Bob's non-selective measurements can be summarized as

$$\frac{\hat{I}+\boldsymbol{s}\hat{\boldsymbol{\sigma}}}{2} \to \frac{\hat{I}+\boldsymbol{s}\hat{\boldsymbol{\sigma}}/3}{2}.$$

6.5 *Anti-linearity of polarization reflection.* Let us calculate the influence of the anti-unitary transformation $\hat{T}$ on a pure state qubit:

$$\hat{T}|\boldsymbol{n}\rangle = \hat{T}\left(\cos\frac{\theta}{2}|\uparrow\rangle + \mathrm{e}^{\mathrm{i}\varphi}\sin\frac{\theta}{2}|\downarrow\rangle\right) = -\cos\frac{\theta}{2}|\downarrow\rangle + \mathrm{e}^{-\mathrm{i}\varphi}\sin\frac{\theta}{2}|\uparrow\rangle = \mathrm{e}^{-\mathrm{i}\varphi}|-\boldsymbol{n}\rangle.$$

6.6 *General qubit effects.* We can suppose that the weights $w_n$ are non-vanishing. First, we have to impose the conditions $|\boldsymbol{a}_n| \leq 1$ since otherwise the matrices would be indefinite. Second, the request $\hat{\Pi}_n \geq 0$ implies the conditions $w_n > 0$. And third, the request $\sum_n \hat{\Pi}_n = \hat{I}$ implies the conditions $\sum_n w_n = 1$ and $\sum_n w_n\boldsymbol{a}_n = 0$.

## Problems of Chap. 7

7.1 *Schmidt orthogonalization theorem.* Let $r$ be the rank of $\hat{c}$ and let us consider the non-negative matrices $\hat{c}\hat{c}^\dagger$ and $\hat{c}^\dagger\hat{c}$ of rank $r$ both. Their spectrum is non-negative and identical. Indeed, if $\hat{c}^\dagger\hat{c}|R\rangle = w|R\rangle$, i.e., $w$ and $|R\rangle$ are an eigenvalue and a (normalized) eigenvector of $\hat{c}^\dagger\hat{c}$ then $|L\rangle = \hat{c}|R\rangle/\sqrt{w}$ will be a (normalized) eigenvector of $\hat{c}\hat{c}^\dagger$ with the same eigenvalue $w$. This can be seen by direct inspection. Now we determine the $r$ positive eigenvalues $w_\lambda$ for $\lambda = 1, 2, \ldots, r$ and the corresponding orthonormal eigenstates $|\lambda; R\rangle$ of $\hat{c}^\dagger\hat{c}$. Then, by $|\lambda; L\rangle = \hat{c}|\lambda; R\rangle/\sqrt{w_\lambda}$, we define the $r$ orthonormal eigenstates of $\hat{c}\hat{c}^\dagger$ which belong to the common positive eigenvalues $w_\lambda$, for $\lambda = 1, 2, \ldots, r$. Now we can see that

$$\hat{c}|\lambda; R\rangle = \sqrt{w_\lambda}|\lambda; L\rangle,$$

for all $\lambda = 1, 2, \ldots, r$. We have thus proven that there exists the following Schmidt decomposition of the matrix $\hat{c}$:

$$\hat{c} = \sum_{\lambda=1}^{r} \sqrt{w_\lambda}|\lambda; L\rangle\langle\lambda; R|.$$

7.2 *Swap operation.* For convenience, we use $\hat{\sigma}_\pm = (\hat{\sigma}_x \pm i\hat{\sigma}_y)/2$ instead of $\hat{\sigma}_x, \hat{\sigma}_y$. Now we express the Pauli matrices in the up-down basis:

$$\hat{\sigma}_+ = |\uparrow\rangle\langle\downarrow|, \quad \hat{\sigma}_- = |\downarrow\rangle\langle\uparrow|, \quad \hat{\sigma}_z = |\uparrow\rangle\langle\uparrow| - |\downarrow\rangle\langle\downarrow|.$$

Substituting these expressions we obtain

$$\begin{aligned}\frac{\hat{I}\otimes\hat{I} + \hat{\boldsymbol{\sigma}}\otimes\hat{\boldsymbol{\sigma}}}{2} &= \frac{\hat{I}\otimes\hat{I} + 2\hat{\sigma}_+\otimes\hat{\sigma}_- + 2\hat{\sigma}_-\otimes\hat{\sigma}_+ + \hat{\sigma}_z\otimes\hat{\sigma}_z}{2} \\ &= |\uparrow\uparrow\rangle\langle\uparrow\uparrow| + |\downarrow\downarrow\rangle\langle\downarrow\downarrow| + |\uparrow\downarrow\rangle\langle\downarrow\uparrow| + |\downarrow\uparrow\rangle\langle\uparrow\downarrow|,\end{aligned}$$

which is indeed the swap matrix $\hat{\mathrm{S}}$.

7.3 *Singlet density matrix.* The singlet state $\hat{\rho}(\text{singlet})$ is invariant under rotations of the Bloch sphere. Therefore $\hat{\rho}(\text{singlet})$ must be of the form

$$\hat{\rho}(\text{singlet}) = \frac{\hat{I}\otimes\hat{I} + \lambda\hat{\boldsymbol{\sigma}}\otimes\hat{\boldsymbol{\sigma}}}{4}$$

because there are no further rotational invariant mathematical structures. We could determine the parameter $\lambda$ from the pure state condition $[\hat{\rho}(\text{singlet})]^2 = \hat{\rho}(\text{singlet})$, yielding $\lambda = -1$. However, we can spare these calculations if we recall the swap $\hat{S}$. It is Hermitian, rotation invariant and idempotent: $\hat{S}^2 = \hat{I}\otimes\hat{I}$. Hence we get the singlet state directly in the form

$$\hat{\rho}(\text{singlet}) = \frac{\hat{I}\otimes\hat{I} - \hat{S}}{2} = \frac{\hat{I}\otimes\hat{I} - \hat{\boldsymbol{\sigma}}\otimes\hat{\boldsymbol{\sigma}}}{4}.$$

7.4 *Local measurement of expectation values.* Alice and Bob will determine $\langle\hat{A}\otimes\hat{B}\rangle$ and $\langle\hat{A}'\otimes\hat{B}'\rangle$ separately on two independent sub-ensembles and will finally add them since the expectation value is additive. Still we have to show that the expectation value of a tensor product, like $\langle\hat{A}\otimes\hat{B}\rangle$, can be determined in local measurements. We introduce the local spectral expansions $\hat{A}=\sum_\lambda A_\lambda\hat{P}_\lambda$ and $\hat{B}=\sum_\mu B_\mu\hat{Q}_\mu$. Alice and Bob perform local measurements of $\hat{A}$ and $\hat{B}$ in coincidence, yielding the measurement outcomes $A_1$, $B_1$, $A_2$, $B_2$, ..., $A_r$, $B_r$, ..., $A_N$, $B_N$ where $A_r$ is always an eigenvalue $A_\lambda$ and the case is similar for the $B_r$'s. Then Alice and Bob can calculate the q-expectation value asymptotically:

$$\langle\hat{A}\otimes\hat{B}\rangle=\lim_{N\to\infty}\frac{1}{N}\sum_r A_rB_r.$$

To see that this is indeed the right expression of $\langle\hat{A}\otimes\hat{B}\rangle$ we have to rethink the nonlocal measurement of $\hat{A}\otimes\hat{B}$ itself. Its spectral expansion is

$$\hat{A}\otimes\hat{B}=\sum_{(\lambda,\mu)}(A_\lambda B_\mu)(\hat{P}_\lambda\otimes\hat{Q}_\mu),$$

and the corresponding q-measurement will obviously yield the same statistics of the outcomes $A_rB_r$ like in case of the local-measurements.

7.5 *Local measurement of certain nonlocal quantities.* If we measure $\hat{\sigma}_z\otimes\hat{\sigma}_z$ on a singlet state we always get $-1$ and the singlet state remains the post-measurement state. In the attempted local measurement, the entanglement is always destroyed and we get either $|\uparrow\downarrow\rangle$ or $|\downarrow\uparrow\rangle$ for the post-measurement state. Obviously the degenerate spectrum of $\hat{\sigma}_z\otimes\hat{\sigma}_z$ plays a role in the nonlocality. If, in the general case, we suppose $\hat{A}\otimes\hat{B}$ has a non-degenerate spectrum then the post-measurement states will be the same pure states in both the nonlocal measurement of $\hat{A}\otimes\hat{B}$ and the simultaneous local measurements of $\hat{A}$ and $\hat{B}$.

7.6 *Nonlocal hidden parameters.* Let the further hidden parameter $v$ take values 1, 2, 3, 4 marking whether Alice and Bob measures $\hat{A}\otimes\hat{B}$, $\hat{A}'\otimes\hat{B}$, $\hat{A}\otimes\hat{B}'$ or $\hat{A}'\otimes\hat{B}'$, respectively. Then, according to the hidden variable concept, the assignment of all four polarization values will uniquely depend on the composite hidden variable $rv$:

$$\hat{A}=A_{rv}=\pm 1,\ \hat{A}'=A'_{rv}=\pm 1,\ \hat{B}=B_{rv}=\pm 1,\ \hat{B}'=B'_{rv}=\pm 1.$$

Contrary to the local assignment (7.36), the above assignments are called nonlocal since the hidden variable $rv$ is nonlocal: it depends on both Alice's and Bob's measurement setup. The statistical relationships, cf. (7.37), become modified:

$$\langle\hat{A}\otimes\hat{B}\rangle=\lim_{N_1\to\infty}\frac{1}{N_1}\sum_{r\in\Omega_1}A_{r1}B_{r1};\ N_1=|\Omega_1|,$$

$$\langle\hat{A}'\otimes\hat{B}\rangle=\lim_{N_2\to\infty}\frac{1}{N_2}\sum_{r\in\Omega_2}A_{r2}B_{r2};\ N_2=|\Omega_2|,$$

etc. for the other two cases $\nu = 3, 4$. The assignments are independent for the four different values of $\nu$. There is no constraint combining the $A_{r\nu}$'s with different $\nu$'s. Hence it has become straightforward to reproduce the above said four q-theoretical predictions including of course correlations $\langle\hat{\mathbf{C}}\rangle$ that are higher than 2.

7.7 *Does teleportation clone the qubit?* The selective post-measurement state of the two qubits on Alice's side is one of the four maximally entangled Bell-states. Therefore the reduced state of the qubit that she had teleported is left in the totally mixed state independently of its original form as well as of the four outcomes of Alice's measurement. Note that the form (7.47) of the three-qubit pre-measurement state shows that Alice's measurement outcome is always random. The four outcomes have probability 1/4 each.

## Problems of Chap. 8

8.1 *All q-operations are reductions of unitary dynamics.* Given the trace-preserving q-operation $\mathcal{M}\hat{\rho} = \sum_n \hat{M}_n\hat{\rho}\hat{M}_n^\dagger$, we have to construct the unitary interaction matrix $\hat{U}$ acting on the composite state of the system and environment. Let us introduce the composite basis $|\lambda\rangle \otimes |n; E\rangle$ where $\lambda = 1, 2, \ldots, d$ and $n = 1, 2, \ldots, d_E$. Let us define the influence of $\hat{U}$ on a subset of the composite basis:

$$\hat{U}(|\lambda\rangle \otimes |1; E\rangle) = \sum_{n=1}^{d_E} \hat{M}_n|\lambda\rangle \otimes |n; E\rangle, \;\; \lambda = 1, 2, \ldots, d.$$

This definition is possible because the above map generates orthonormal vectors:

$$\sum_{m=1}^{d_E} \langle\mu|\hat{M}_m^\dagger \otimes \langle m; E| \sum_{n=1}^{d_E} \hat{M}_n|\lambda\rangle \otimes |n; E\rangle = \sum_{n=1}^{d_E} \langle\mu|\hat{M}_n^\dagger\hat{M}_n|\lambda\rangle = \delta_{\lambda\mu}.$$

The further matrix elements of $\hat{U}$, i.e. those not defined by our first equation above, can be chosen in such a way that $\hat{U}$ is unitary on the whole composite state. Using this definition of $\hat{U}$ in the equation (8.3) of reduced dynamics we can directly inspect that the resulting operation is $\mathcal{M}\hat{\rho} = \sum_n \hat{M}_n\hat{\rho}\hat{M}_n^\dagger$, as expected.

8.2 *Non-projective effect as averaged projection.* Let us substitute the proposed form of the effects $\hat{\Pi}_n$ into the equation $p_n = \mathrm{tr}(\hat{\Pi}_n\hat{\rho})$ introduced for non-projective measurement in Sect. 4.4.2

$$\mathrm{tr}(\hat{\Pi}_n\hat{\rho}) = \mathrm{tr}(\mathrm{tr}_E\hat{P}_n\hat{\rho}_E\hat{\rho}) = \mathrm{tr}(\hat{P}_n\hat{\rho}_E\hat{\rho}).$$

In this formalism, i.e., without the $\otimes$'s, the matrices of different subsystems commute hence $\hat{\rho}_E\hat{\rho} = \hat{\rho}\hat{\rho}_E$. Thus we obtain the following result: $\mathrm{tr}(\hat{\Pi}_n\hat{\rho}) = \mathrm{tr}(\hat{P}_n\hat{\rho}\hat{\rho}_E)$. We recognize the coincidence of the r.h.s. with the r.h.s. of (8.18). Since this coincidence is valid for all possible $\hat{\rho}$, it verifies the proposed form of $\hat{\Pi}_n$.

8.3 *Q-operation as supermatrix.* We start from the Kraus representation $\mathcal{M}\hat{\rho} = \sum_n \hat{M}_n\hat{\rho}\hat{M}_n^\dagger$. We take the matrix elements of both sides and we also sandwich the $\hat{\rho}$ between the identities $\sum_{\lambda'}|\lambda'\rangle\langle\lambda'|$ and $\sum_{\mu'}|\mu'\rangle\langle\mu'|$ on the r.h.s.:

$$\langle\lambda|\mathcal{M}\hat{\rho}|\mu\rangle = \langle\lambda|\sum_n \hat{M}_n \sum_{\lambda'}|\lambda'\rangle\langle\lambda'|\hat{\rho}\sum_{\mu'}|\mu'\rangle\langle\mu'|\hat{M}_n^\dagger|\mu\rangle.$$

Comparing the r.h.s. with $\sum_{\lambda'\mu'}\mathcal{M}_{\lambda\mu\lambda'\mu'}\rho_{\lambda'\mu'}$, we read out the components of the supermatrix: $\mathcal{M}_{\lambda\mu\lambda'\mu'} = \sum_n\langle\lambda|\hat{M}_n|\lambda'\rangle\langle\mu'|\hat{M}_n^\dagger|\mu\rangle$.

8.4 *Environmental decoherence, time-continuous depolarization.* The equation takes the Lindblad form with $\hat{H}=0$ and with three hermitian Lindblad matrices identified by the Cartesian components of $(\hat{\boldsymbol{\sigma}}/2\sqrt{\tau})$. For convenience, we shall use the Einstein convention to sum over repeated indices, e.g.: $\boldsymbol{s}\hat{\boldsymbol{\sigma}} = s_a\hat{\sigma}_a$. We write the r.h.s. of the master equation into the equivalent form $-(1/8\tau)[\hat{\sigma}_b,[\hat{\sigma}_b,\hat{\rho}]]$ and insert $\hat{\rho} = \frac{1}{2}(\hat{I}+s_a\hat{\sigma}_a)$ into it. The master equation reduces to

$$\dot{s}_a\hat{\sigma}_a = -\frac{1}{8\tau}[\hat{\sigma}_b,[\hat{\sigma}_b,s_a\hat{\sigma}_a]] = -\frac{1}{\tau}s_a\hat{\sigma}_a,$$

which means the simple equation $\dot{\boldsymbol{s}} = -\boldsymbol{s}/\tau$ for the polarization vector. Its solution is $\boldsymbol{s}(t) = \mathrm{e}^{-t/\tau}\boldsymbol{s}(0)$. Therefore the $\tau$ may be called depolarization time, or decoherence time as well.

8.5 *Kraus representation of depolarization.* The map should be of the form $\mathcal{M}\hat{\rho} = (1-3\lambda)\hat{\rho} + \lambda\hat{\boldsymbol{\sigma}}\hat{\rho}\hat{\boldsymbol{\sigma}}$ with $0\le\lambda\le 1/3$ since there exist no other isotropic Kraus structures for a qubit. The depolarization channel decreases the polarization vector $\boldsymbol{s}$ by a factor $1-w$ and we have to find the parameter $\lambda$ as function of $w$. Inserting $\hat{\rho} = \frac{1}{2}(\hat{I}+\boldsymbol{s}\hat{\boldsymbol{\sigma}})$ we get

$$\mathcal{M}\frac{\hat{I}+\boldsymbol{s}\hat{\boldsymbol{\sigma}}}{2} = \frac{\hat{I}+(1-4\lambda)\boldsymbol{s}\hat{\boldsymbol{\sigma}}}{2},$$

which means that $\lambda = w/4$. Four Kraus matrices make the depolarization channel: $\sqrt{1-3w/4}\hat{I}$ and the three components of $\sqrt{w/4}\hat{\boldsymbol{\sigma}}$.

## Problems of Chap. 9

9.1 *Positivity of relative entropy.* We can write

$$S(\rho\|\rho') = \sum_{\mathrm{x}} \rho(\mathrm{x}) \log \frac{\rho(\mathrm{x})}{\rho'(\mathrm{x})}.$$

We invoke the inequality $\ln \lambda > 1 - \lambda^{-1}$ valid for $\lambda \neq 1$ and apply it to $\lambda = \rho/\rho'$. This yields

$$S(\rho\|\rho') > \frac{1}{\ln 2} \sum_{\mathrm{x}} \rho(\mathrm{x}) \left[1 - \frac{\rho'(\mathrm{x})}{\rho(\mathrm{x})}\right] = 0,$$

which always holds if $\rho' \neq \rho$.

9.2 *Concavity of entropy.* Suppose we have a long message $\mathrm{x}_1^{(1)}\mathrm{x}_2^{(1)}\ldots\mathrm{x}_n^{(1)}$ where $\rho_1(\mathrm{x})$ is the a priori distribution of one letter. Let $S_1$ stand for the single-letter entropy $S(\rho_1)$. The number of the typical messages is $2^{nS_1}$ so that their shortest code is $nS_1$ bits. Consider a second message from the same alphabet and suppose the single-letter distribution $\rho_2(\mathrm{x})$ is different from $\rho_1(\mathrm{x})$. Let us concatenate the two messages:

$$\mathrm{x}_1^{(1)}\mathrm{x}_2^{(1)}\mathrm{x}_3^{(1)}\ldots\mathrm{x}_n^{(1)}\mathrm{x}_1^{(2)}\mathrm{x}_2^{(2)}\mathrm{x}_3^{(2)}\ldots\mathrm{x}_m^{(2)},$$

where the two lengths $n$ and $m$ may be different. Obviously, the number of the typical ones among such composite messages is $2^{nS_1} \times 2^{mS_2}$ and their shortest code is $nS_1 + mS_2$ bits. Now imagine that we permute the $n + m$ letters randomly. On one hand, the composite messages become usual $(n + m)$-letter-long messages where the single letter distribution is always the same, i.e., the mixture $\rho = w_1\rho_1 + w_2\rho_2$ with weights $w_1 = n/(n + m)$ and $w_2 = m/(n + m)$. Therefore the number of the typical messages is $2^{(n+m)S(\rho)}$ and the shortest code is $(n + m)S(\rho)$ bits. On the other hand, we can inspect that the number of the typical messages $2^{(n+m)S(\rho)}$ is greater than $2^{nS_1} \times 2^{mS_2}$ because the number of inequivalent permutations has increased: the first $n$ letters have become permutable with the last $m$ letters. This means that $(n + m)S(\rho) > nS_1 + mS_2$ which is just the concavity of the entropy: $S(w_1\rho_1 + w_2\rho_2) > w_1S(\rho_1) + w_2S(\rho_2)$, for $\rho_1 \neq \rho_2$.

9.3 *Subadditivity of entropy.* We can make the choice $\rho'_{AB}(\mathrm{x},\mathrm{y}) = \rho_A(\mathrm{x})\rho_B(\mathrm{y})$. To calculate $S(\rho_{AB}\|\rho'_{AB}) = -S(\rho_{AB}) - \sum_{\mathrm{x,y}} \rho_{AB}(\mathrm{x},\mathrm{y}) \log \rho'_{AB}(\mathrm{x},\mathrm{y})$, we note that the second term is $-\sum_{\mathrm{x,y}} \rho_{AB}(\mathrm{x},\mathrm{y}) \log[\rho_A(\mathrm{x})\rho_B(\mathrm{y})] = S(\hat\rho_A) + S(\hat\rho_B)$. Hence the positivity of the relative entropy $S(\rho_{AB}\|\rho'_{AB}) \geq 0$ proves subadditivity: $S(\rho_{AB}) \leq S(\rho_A) + S(\rho_B)$.

9.4 *Coarse graining increases entropy.* Let us identify our system by the $k$-partite composite system of the $k$ bits $\mathrm{x}_1, \mathrm{x}_2, \ldots, \mathrm{x}_k$. Then the coarse grained system corresponds to the $(k - 1)$-partite sub-system consisting of the first $k - 1$ bits $\mathrm{x}_1, \mathrm{x}_2, \ldots, \mathrm{x}_{k-1}$. The coarse grained state $\tilde\rho$ is just a reduced state w.r.t. $\rho$. Hence we see that coarse graining increases the entropy because reduction does it.

## Problems of Chap. 10

10.1 *Subadditivity of q-entropy.* Let us calculate

$$S(\hat{\rho}_{AB}\|\hat{\rho}_A \otimes \hat{\rho}_B) = -S(\hat{\rho}_{AB}) - \mathrm{tr}[\hat{\rho}_{AB}\log(\hat{\rho}_A \otimes \hat{\rho}_B)]$$

and note that the second term is $S(\hat{\rho}_A) + S(\hat{\rho}_B)$. Hence the Klein inequality $S(\hat{\rho}'\|\hat{\rho}) \geq 0$ proves the subadditivity: $S(\hat{\rho}_{AB}) \leq S(\hat{\rho}_A) + S(\hat{\rho}_B)$.

10.2 *Concavity of q-entropy, Holevo entropy.* We assume a certain environmental system $E$ and a basis $\{|n;\, E\rangle\}$ for it. Let us construct a composite state

$$\hat{\rho}_{big} = \sum_n w_n \hat{\rho}_n \otimes |n; E\rangle\langle n; E|.$$

Note that the reduced state of the system is invariably $\hat{\rho} = \sum_n w_n \hat{\rho}_n$ and the reduced state of the environment is $\hat{\rho}_E = \sum_n w_n |n; E\rangle\langle n; E|$. Subadditivity guarantees that $S(\hat{\rho}_{big}) \leq S(\hat{\rho}) + S(\hat{\rho}_E)$. Let us calculate and insert the entropies $S(\hat{\rho}_{big}) = S(w) + \sum_n w_n S(\hat{\rho}_n)$ and $S(\hat{\rho}_E) = S(w)$ which results in the desired inequality: $\sum_n w_n S(\hat{\rho}_n) \leq S(\hat{\rho})$.

10.3 *Data compression of the non-orthogonal code.* The density matrix of the corresponding 1-letter q-message reads

$$\hat{\rho} = \frac{|\uparrow z\rangle\langle\uparrow z| + |\uparrow x\rangle\langle\uparrow x|}{2} = \frac{\hat{I} + \hat{\sigma}_n/\sqrt{2}}{2},$$

which is a partially polarized state along the skew direction $n = (1,0,1)/\sqrt{2}$, cf. (6.29). The eigenvalues of this density matrix are the following:

$$p_+ = \frac{1 + 1/\sqrt{2}}{2}, \quad p_- = \frac{1 - 1/\sqrt{2}}{2},$$

hence its von Neumann entropy amounts to

$$S(\hat{\rho}) = -p_+ \log p_+ - p_- \log p_- \approx 0.60.$$

According to the q-data compression theorem, we can compress one qubit of the q-message into 0.6 qubit on average, and this is the best maximum faithful compression.

10.4 *Distinguishing two non-orthogonal qubits: various aspects.* In fact, we measure the polarization component orthogonal to the polarization of the single-letter density matrix. The measurement outcomes $\pm 1$ on both q-states $|\uparrow z\rangle, |\uparrow x\rangle$ will appear with probabilities $p_+$ and $p_-$ (cf. Prob. 10.3), in alternating order of course:

$$p(y=+1|x=0)=p_+,\quad p(y=-1|x=0)=p_-$$
$$p(y=+1|x=1)=p_-,\quad p(y=-1|x=1)=p_+.$$

Regarding the randomness of the input message the output message, too, becomes random: $H(Y)=1$. Hence the information gain takes this form and value:

$$I_{gain}=H(Y)-H(Y|X)=1+p_+\log p_+ + p_-\log p_- \approx 0.40.$$

10.5 *Simple optimum q-code.* The q-data compression theory says that a pure state q-code is not compressible faithfully (i.e.: allowing the same accessible information) if and only if the single-letter average state has the maximum von Neumann entropy. In our case, we must assure the following:

$$\frac{|R\rangle\langle R|+|G\rangle\langle G|+|B\rangle\langle B|}{3}=\frac{\hat{I}}{2},$$

which is possible if we chose three points on a main circle of the Bloch-sphere, at equal distances from each other.

## Problems of Chap. 11

11.1 *Creating the totally symmetric state.*

$$|S\rangle \equiv \frac{1}{2^{n/2}}\sum_{x=1}^{2^n-1}|x_1x_2...x_n\rangle=\sum_{x_1=0}^{1}\sum_{x_2=0}^{1}\cdots\sum_{x_n=0}^{1}|x_1\rangle\otimes|x_2\rangle\otimes\ldots\otimes|x_n\rangle$$
$$=\hat{H}|0\rangle\otimes\hat{H}|0\rangle\otimes\ldots\otimes\hat{H}|0\rangle\equiv\hat{H}^{\otimes n}|0\rangle^{\otimes n}.$$

11.2 *Constructing Z-gate from X-gate.*

$$\hat{H}\hat{X}\hat{H}=\frac{1}{\sqrt{2}}\begin{bmatrix}1 & 1\\ 1 & -1\end{bmatrix}\begin{bmatrix}0 & 1\\ 1 & 0\end{bmatrix}\frac{1}{\sqrt{2}}\begin{bmatrix}1 & 1\\ 1 & -1\end{bmatrix}=\begin{bmatrix}1 & 0\\ 0 & -1\end{bmatrix}=\hat{Z}.$$

The inverse relationship follows from $\hat{H}^2=\hat{I}$.

11.3 *Constructing controlled Z-gate from cNOT-gate.*

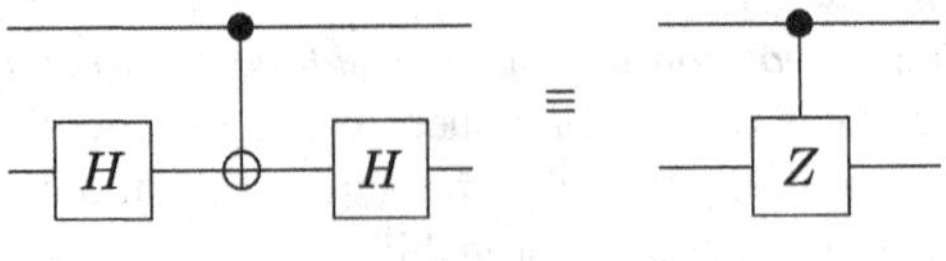

11.4 *Constructing controlled phase-gate from two cNOT-gates.*

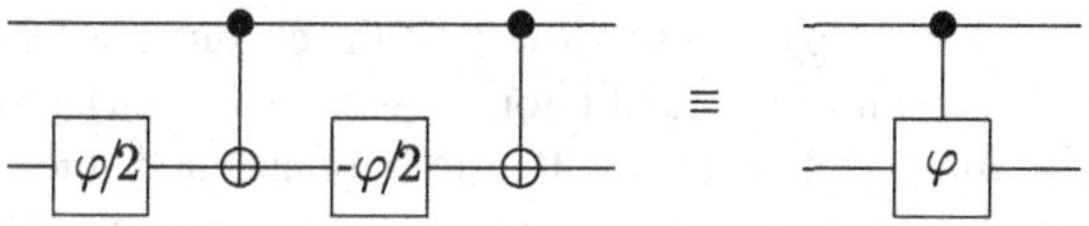


11.5 *Q-circuit to produce Bell states.* Let us calculate the successive actions of the H-gate and the cNOT-gate:

$$\begin{array}{lllllll} |00\rangle & \longrightarrow & |00\rangle + |10\rangle & \longrightarrow & |00\rangle + |11\rangle & \longrightarrow & |\Phi^{(+)}\rangle \\ |01\rangle & \longrightarrow & |01\rangle + |11\rangle & \longrightarrow & |01\rangle + |10\rangle & \longrightarrow & |\Psi^{(+)}\rangle \\ |10\rangle & \longrightarrow & |00\rangle - |10\rangle & \longrightarrow & |00\rangle - |11\rangle & \longrightarrow & |\Phi^{(-)}\rangle \\ |11\rangle & \longrightarrow & |01\rangle - |11\rangle & \longrightarrow & |01\rangle - |10\rangle & \longrightarrow & |\Psi^{(-)}\rangle \end{array}$$

The trivial factors $1/\sqrt{2}$ in front of the intermediate states have not been denoted.

11.6 *Q-circuit to measure Bell states.* The task is the inverse task of preparing the Bell states. Since both the H-gate and the cNOT-gate are the inverses of themselves, respectively, we can simply use them in the reversed order w.r.t. the circuit that prepared the Bell states (cf. Prob. 11.5):

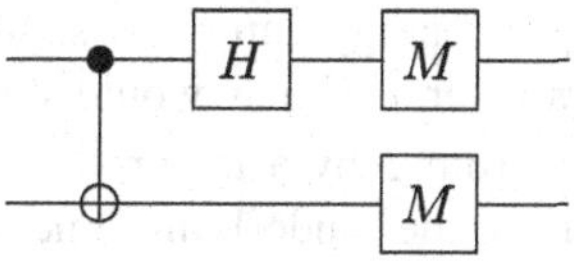


The boxes $\boxed{M}$ stand for projective measurement of the computational basis.

## Problems of Chap. 12

12.1 *Terabyte equivalent to Joule/degree.* To heat up 1 g water by 1 K we must deliver 4.186 J heat to it. If we adopt 300 K for room temperature, the increase of the thermodynamic entropy will be $S_{\text{th}} = (4.186/300)$ J/K. Using $k_B = 1.381 \times 10^{-23}$ J/K, our Eq. (12.11) yields $S \sim 10^{21}$ bits for the increase of the informatic entropy. One terabyte is just $8 \times 10^{12}$ bits, it corresponds to an amount of water less than 1 g by cca. 9 orders of magnitude. We conclude that the rate 1 terabyte/degree of informatic entropy increase would correspond to 1 ng water—a drop of size $\sim$ 10 μm.

12.2 *Landauer's principle.* Let us consider a large q-storage consisting of $N$ qubits and suppose, for concreteness, that they contain maximally compressed data. The von Neumann entropy is $S = N$. Suppose that the environment of the q-storage is thermal, its temperature is $T$, and it is uncorrelated with the q-storage. When the storage becomes erased, i.e., all qubits are set to $|0\rangle$, the von Neumann entropy drops to $S = 0$. If the erasure has been done reversibly from the viewpont of the total (storage + environment) system then the total q-entropy must not change. If the storage and the environment are uncorrelated after the erasure, too, then the change $-N$ of the q-storage entropy will be compensated by the increase $Nk_B\ln 2$ of the thermodynamic entropy of the environment. This is achieved by dissipating work $Nk_BT\ln 2$ into heat in the environment. The dissipated heat is, indeed, $k_BT\ln 2$ per one erased qubit. If the erasure were not reversible, the released heat would be greater.

12.3 *Universal thermalizer?* To solve Eq. (12.16) with the Hamiltonian $\hat{H}' = \epsilon'\hat{n}$, we use the common interaction picture of quantum theory. This means the replacements $\hat{a} \rightarrow \exp(-i\epsilon' t/\hbar)\hat{a}$ and $\hat{\rho}_t \rightarrow \exp(i\epsilon'\hat{n}t/\hbar)\hat{\rho}_t\exp(-i\epsilon'\hat{n}t/\hbar)$. The master equation becomes

$$\frac{d\hat{\rho}}{dt} = \Gamma\left(\hat{a}\hat{\rho}_t\hat{a}^\dagger - \frac{1}{2}\{\hat{a}^\dagger\hat{a}, \hat{\rho}_t\}\right) + e^{-\beta\epsilon}\Gamma\left(\hat{a}^\dagger\hat{\rho}_t\hat{a} - \frac{1}{2}\{\hat{a}\hat{a}^\dagger, \hat{\rho}_t\}\right).$$

The self-Hamiltonian term cancels—this is always so in interaction picture. Also the time dependence of $\hat{a}$ and $\hat{a}^\dagger$ cancel—this is a special consequence of the Lindblad structure. Note that the above obtained form is not sensitive to the value of the qubit energy gap $\epsilon'$. Its stationary solution will invariably be $\hat{\rho}_\beta$ with the parameter $\epsilon$ which would correspond to a Gibbs state at temperature different from $T$ by a factor $\epsilon'/\epsilon$. This incapacity of the master equation follows from the underlying microscopic model. The thermal reservoir $\hat{\rho}_\beta^{\otimes N}$ thermalizes to the same (inverse) temperature $\beta$ provided the qubit to be thermalized has exactly the same Hamiltonian as the reservoir's qubits have. This is not the case here.

12.4 *Mechanical work on a qubit.* Suppose the vertical magnetic field $\omega$ depends on the $z$-coordinate. The mean energy $E = \mathrm{tr}(\hat{H}\hat{\rho})$ of the isolated qubit behaves like a $z$-dependent potential energy. To keep the qubit at a certain vertical location $z$ one must exert a mechanical force $dE/dz$. Now, one starts to vary the external magnetic field just by moving the qubit along the vertical axis. This requires mechanical work $W$ against the magnetic field gradient. The requested power is $(dE/dz)(dz/dt) = \mathrm{tr}\left[(d\hat{H}/dz)\hat{\rho}\right](dz/dt) = \mathrm{tr}\left[(d\hat{H}/dt)\hat{\rho}\right]$ which coincides with the definition (12.24) of the power $dW/dt$.

12.5 *Adiabatic demagnetization.* The thermalized state of the qubit in the field $\omega$ is $\hat{\rho}_{\beta,\omega} = \mathcal{N}\exp(\beta\hbar\omega\hat{\sigma}_z)$. If we change the field quickly to $\omega'$, the state remains as it was. The thermal environment will drive it to the new thermal state $\hat{\rho}_{\beta,\omega'} = \mathcal{N}\exp(\beta\hbar\omega'\hat{\sigma}_z)$. But it takes time and right after the quick transition $\omega \rightarrow \omega'$ the qubit is surely of the previous equilibrium state $\hat{\rho}_{\beta,\omega}$.

Now we observe that this state, at the current (smaller) field, corresponds to an other thermal state

$$\hat{\rho}_{\beta,\omega} = \mathcal{N} \exp(\beta\hbar\omega\hat{\sigma}_z) = \mathcal{N} \exp(\beta'\hbar\omega'\hat{\sigma}_z) = \hat{\rho}_{\beta',\omega'}$$

where $\beta' = (\omega/\omega')\beta$. Hence the effective temperature $T' = \beta'/k_B$ of our qubit is

$$T' = \frac{\omega'}{\omega}T \ll T.$$

This means that, during the process of re-thermalization to temperature $T$, our qubit absorbes heat from the environment, it acts as a refrigerator.

# Index


**A**
Alice, Bob, 41, 51, 63, 66, 67, 69, 71, 72
Alice, Bob, Eve, 55

**B**
Bayes theorem, 8
Bell
basis, states, 65
inequality, 70
nonlocality, 68
bit, 13

**C**
channel capacity, 91
code, 88, 123
optimum, 92
q-, 96
superdense, 71
completely positive map, 23, 77
contrary to classical, 22, 23, 26, 34

**D**
data compression, 88, 97, 123
decoherence, 26
density matrix, 22, 38

**E**
entanglement, 21, 82
as resource, 100
dilution, 102
distillation, 64, 101
maximum, 65
measure, 61
two-qubit, 64
entropy
conditional, 91
informatic, 123, 135
relative, 88, 96, 138
Shannon, 62, 87
thermodynamic, 123, 135
von Neumann, 62, 95, 135
equation
master, 29, 83, 125
of motion, 5, 21, 39
expectation value, 9, 27

**F**
fidelity, 40, 51
Fock representation, 42, 124
function evaluation, 108

**I**
information, q-information, 13, 95
accessible, 99
mutual, 90
theory, 87, 95
irreversible
master equation, 29
operation, 12, 33, 47, 138
q-measurement, 26
reduced dynamics, 83

**K**
Klein inequality, 76
Kraus form, 77

**L**
Lindblad form, 83, 125
local
  Hamilton, 63
  operation, 63
  physical quantity, 63
LOCC, 82

**M**
measurement, 8, 24, 40
  continuous, 10, 27
  in pure state, 30
  indirect, 79, 81
  non-projective, 10, 21, 27, 52, 81
  non-selective, 10, 26
  projective, 8, 25, 52
  selective, 9, 25
  unsharp, 10
  weak, 10, 27
message, 88, 96
  typical, 89
mixing, 6, 22, 48, 49

**N**
nonlocality
  Bell, 68
  Einstein, 66

**O**
operation, 6, 22, 77
  depolarization, 47
  irreversible
  local, 63
  logical, 63
  non-selective, 4, 23
  one-qubit, 45
  reflection, 47
  selective, 7, 23

**P**
Pauli, 35
  matrices, 37
  representation, 35, 41
physical quantities, 8, 24
  compatible, 29

**Q**
q-algorithm
  error correction, 116
  Fourier, 113
  oracle problem, 109
  period finding, 114
  searching, 111
q-banknote, 54
q-Carnot cycle, 128
q-channel, 56, 83
q-circuit, 118
q-computation, 103
  parallel, 103
  representation
q-correlation, 71, 32
  history, 66
q-cryptography, 55
q-entropy, 95
q-gate, 118
  universal, 47
q-information
  hidden
q-key, 55
q-protocol, 54
q-refrigerator, 125
q-state
  cloning, 51
  determination, 48, 51
  indistinguishability, 51, 54
  no-cloning, 50, 54
  non-orthogonal, 51
  preparation, 48
  purification
  unknown, 45, 49
q-thermalization
qubit, 18, 35
  external work, 127
  ideal gas, 122
  thermal, 121
  unknown, 41

**R**
reduced dynamics, 12, 78
reservoir
rotational invariance

**S**
Schmidt decomposition
selection, 6, 22
state space, 5, 21
  discrete, 14
state, q-state
  mixed, 5, 21
  pure, 5, 21
  separable, 12

superposition, 18, 21
system
  bipartite, 59
  collective, 13
  composite, 12
  environmental, 60
  open, 86

**T**
teleportation, 72

**U**
urn model, 5–7, 22, 24